ALICE
AND THE
QUANTUM
CAT

WILLIAM B. SHANLEY, EDITOR
NICK HERBERT, SCIENCE EDITOR

alice
and the
Quantum
cat

ORIGINAL CONCEPT BY
DIANNE COLLINS, WILLIAM B. SHANLEY
AND FRED ALAN WOLF

Publishing

A catalogue record for this book is available
from the British Library

ISBN 978-88-95604-10-7

Printed and bound in the United States

Book and cover design by Andrea Barbieri
Cover illustration by Mie Katrine Brieghel

Pari Publishing

Via Tozzi 7, 58045 Pari (GR), Italy
www.paripublishing.com

TABLE OF CONTENTS

PART II: QUANTUM SPECULATIONS & JOURNEYS TO THE OTHER SIDE

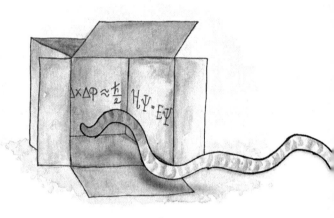

THE QUANTUM AGE[1]

William Brandon Shanley

> There may be no such thing as the glittering central mechanism
> of the universe. Not machinery but magic may be the better de-
> scription of the treasure that is waiting.
> JOHN A. WHEELER, PHYSICIST

A revolution in thought is taking place all around us, a revolution beyond our senses. It is a change so profound, and so far-reaching, that it
promises to be the most momentous since Sir Isaac Newton's discovery of
the 'Laws of Nature' changed our thinking from mystical to mechanical
and ushered in the Age of Reason.

The great leap in scientific knowledge occurred in 1900 in Berlin when
Max Planck proposed the existence of the quantum. Energy is emitted in
quanta, or discrete bursts of energy not in a continuous flow as had previously been thought. Moreover quantum (or subatomic) particles engaged in
discrete jumps between energy states in the orbits of atoms without appearing to move between—the famous quantum jump. Taken together these
anomalies meant that Newton's theory no longer held at the foundations
of reality 'An act of desperation...a theoretical explanation had to be sup-
plied at all cost, whatever the price,' said Planck.[2] And who could have fore-
seen the most remarkable discovery of all—the Observer Effect? Who could
have known that by reaching out and touching the universe we would be
disturbing it—not only here—but far away?

Ever since the universe jumped and we leaped into the Quantum Age,
our concept of reality has been transforming. All matter in the universe is
now seen as instantly connected in some mysterious way; in a universe of
energy the fundamental states of matter exist only as a possibility until they
are observed; and, upon observation, the form they take obligingly depends
on what they are asked to do!

1. Thanks to Fred Alan Wolf, Danah Zohar, Huston Smith and Amit Goswami.
2. Fred Alan Wolf, *Taking the Quantum Leap,* Harper & Row, New York: 1981.
p. 58.

Newtonian physics implied the universe was a vast machine—the quantum model showed there is no machine, but a mysterious entanglement with the observer. The area of preparation must now include the participant observer. Newtonian physics suggested an end to free will and creativity—the quantum model put the observer back into the universe as a participant/creator.

Newton's universe was seen as a giant clockwork made of energy/matter which God wound up at the beginning of time and left running unattended. Events developed in an orderly, continuous fashion according to Newton's laws, just as the planets circle the Sun. Knowing Nature's mechanical laws plus the conditions at the beginning of the universe, man could, in principle, predict every future event.

In Newton's inexorable machine, the universe was made of dead, insensate matter and isolated separate parts, acting in opposition, like so many billiard balls. In the quantum model, we see the universe as an amazingly complicated, finely balanced assemblage of forces and particles woven together in exquisite detail dancing in space-time. In sharp contrast with Newton's model, Quantum Reality is possibilistic and probabilistic, not fixed and determined. Energy seeks to fulfill all potentials, all at once, in every possible variation. The Quantum Universe is holistic—alive, creative, interconnected, interpenetrated, and communicating with itself at every level of Creation. In this livelier, subtle universe, consciousness plays an essential role.

Technology, cities, economies, wars, riots, crime and environmental destruction are the result, in the end, of a way of thinking. The machine metaphor ruthlessly foreclosed the finer attributes and meanings of Creation—Mind, Spirit and Life's purposive action to become. Newton's mechanical, material worldview provided the foundation for Marx's deterministic laws of history, Darwin's mechanistic, reductionist theory of blind evolutionary struggle, and Freud's 'scientific' model of the self as a complex hydraulic system. The mechanistic universe model continues to be the central paradigm of the world today, even more than a century after Planck's discovery.

Civilization's over reliance on classical physics, empiricism and the objectivity of the Scientific Method, has ushered French mathematician Pierre Simon La Place's frightening 'nightmare of determinism' imprisoning humanity in the gears of Newton's unfeeling machine. 'The world that science presents to our belief,' philosopher Bertrand Russell wrote at the turn of the 20th century tells us,

> ...that man is the product of causes which had no provision
> of the end they were achieving; that his origin, his growth, his
> hopes and fears, his loves and his beliefs, are but the outcome
> of accidental collocations of atoms; that no fire, no heroism, no
> intensity of thought and feeling, can preserve the individual life
> beyond the grave; that all the labors of the ages, all the devotion,

all the inspiration, all the noonday brightness of human genius,
are destined to extinction in the vast death of the solar system,
and that the whole temple of Man's achievement must inevitably
be buried beneath the debris of a universe in ruins...

As we look out into the world around us, we see corporations and
governments all too frequently forcing life into strict machine-like time/
space frameworks for the purpose of predictability, control and short-term
profit. Since a machine's only meaning is its purpose, and if a human being is nothing more than a replaceable part in a mechanistic paradigm of
power, predictability and control, a meaningful life is impossible. As such,
why do we continue to cling to this insensitive, and outmoded, paradigm
in our daily lives?

The best answer may be for money. The mechanistic model has made
possible the ability to manufacture products on time, on budget, with un-
erring predictability, making possible precise determinations of overhead
and profit. These projections have unleashed swarms of Ivy-leagued she-
wolves of Wall Street on communities around the world, armed with their
calculators measuring quantities of things, but who all too often trample
the unquantifiable qualitative distinctions of life like so many skilled bar-
barians seeking plunder.

In the wake of this materialistic void, in this universe without meaning,
we are destined to swim in a sea of unprecedented social malaise, psychological dysfunction, addiction and violence—for we feel powerless.

At its most fundamental level, the diagnosis of alienation is based
on the view that modernization forces upon us a world that,
although baptized as real by science, is denuded of all humanly
recognizable qualities...the scientific world view makes it illegiti-
mate to speak of them as being 'objectively' part of the world,
forcing us instead to define such evaluation and such emotional
experience as 'merely subjective' projections of people's inner
lives. The world, once an 'enchanted garden,' to use Max Weber's
memorable phrase, has now become disenchanted, deprived of
purpose and direction, bereft—in these senses—of life itself. All
that is allegedly basic to the specifically human status in nature
comes to be forced back upon the precincts of the 'subjective'
which, in turn, is pushed by the modern scientific view ever more
into the province of dreams and illusions.
MANFRED STANLEY, PHILOSOPHER[3]

3. Manfred Stanley, 'Beyond Progress: Three Post-Political Futures,' from Huston
Smith, *Beyond the Post-Modern Mind*, Crossroad, New York: 1982. p. 84.

But the ground is shaking beneath our feet. The machine is sinking in the quicksand of scientific revolution. Old answers and familiar ways of doing things based on the machine metaphor simply aren't working anymore. Our institutions are in free-fall, gridlock, and disintegration from within. It seems that no number of new laws or methods of enforcement can stem this tide. Too many anomalies contradict official stories for the centralized structures of governmental power to hold. In the words of futurist Alvin Toffler, 'The Age of the Machine is screeching to a halt.'

The level of thinking that has brought us to this juncture is woefully inadequate to resolve the crises of meaning we now face. We are searching for nothing less than an entirely new way of seeing, thinking and being. As Albert Einstein said, 'The world that we have made as a result of the level of thinking we have done thus far, creates problems that we cannot solve at the same level we created them.' The million new faces born each day demand no less of us.

We know Newton's model works for large objects in the familiar, day-to-day world, but at the most fundamental level of reality, that of atoms and quanta, any attempt to verify Newton's model disrupts the very thing we are looking for. We have been looking at the world and each other through a dark, distorted, diminutive conceptual lens. Ironically, the real illusion is the familiar world of our senses, for science now tells us that we live our lives in the thinnest of veneers, as little as one percent of what it can now measure and calculate to exist. And yet, we insist on living our lives in this paradigm of gross materialism that obeys only Newton's laws.

In Newton's world, ambiguity was the enemy—mechanism stresses the absolute, the unchanging and the certain—things are 'either/or,' 'good/bad.' In the quantum world reality is 'both/and'—a coexistence of mutually contradictory possibilities, all equally true, each one a potentially possible constituent of reality. Acausal, non-local synchronicities can give rise to events that seem to 'pop-up' out of thin air. There are no isolated, separate, closed systems in Nature. In this universe of wholeness, everything affects everything else, from the most fundamental particles to faraway galaxies at the edge of the universe.

The objective observer, the very basis of the Scientific Method and the hallmark of our political, judicial and journalistic systems—must now give way to the 'Observer Effect': Instead of seeing events only in terms of the causes and effects of isolated separate parts, we now must also look at our events in terms of context, causal weaves and relationships. This also means we can't control things like we thought we could. In a sense we need to let go of force as the answer and dance with possibilities for inspired action aligned with new visions of the future.

As Danah Zohar has written, '...the details of quantum physics, and the sweeping conceptual revolution that underpins it, have made almost no impact on our perceptions of ourselves or the world around us. We simply don't understand it, and most of us, like Alice, think we can't.'

What we are suggesting is that we are entering a revolution in thought and technology more astonishing than anything we could ever imagine. What's more, this revolution may be our one hope of saving the planet and healing the wounds of humanity.

And just what does this revolution with its new ways of seeing, new meanings and new ways of being bring to our awareness about the nature of reality? To Amit Goswami reality is an epiphenomenon of consciousness, not the other way around.

> If people only knew that at the *most fundamental level*, we are not made of atoms—and that things are not made of atoms—but the consciousness is the ground of being—and that we create our reality literally—then our behavior, our relationship with every-thing in our experience and in our environment would undergo radical change.
> AMIT GOSWAMI, PHYSICIST[4]

The inevitable revolution in thought that might have been expected to follow the discovery of the quantum nature of reality has been slow in coming. This may be due to the fact that, unlike Newtonian physics that describes our everyday world, quantum physics describes the unimaginably small, subtle and counter-intuitive that underpins it. Be that as it may, the revolution is nigh.

> Surprisingly, many features of the new physical reality mirror, un-cannily, qualities some progressive thinkers are hoping to evoke in a new social reality: holistic, beyond dichotomy, plural, respon-sive, emergent, 'green,' spiritual, and in dialogue with science. [5]

> The idea of a quantum society stems from the conviction that a whole new paradigm is emerging from our description of quan-tum reality and that this paradigm can be extended to change radically our perception of ourselves and the social world we want to live in. I believe that a wider appreciation of the revolu-tionary nature of quantum reality, and the possible links between

4. From a 1995 interview with William Brandon Shanley.

5. Danah Zohar, *The Quantum Society*. William Morrow, New York: 1994. pp. 29-32.

quantum processes and our own brain processes, can give us the conceptual foundations we need to bring about a positive revolution in society.

DANAH ZOHAR [6]

A TWENTY-FIRST CENTURY MYTH

'The time has come,' the Walrus said,
'To talk of many things:
Of shoes—and ships—and sealing-wax—
Of cabbages—and kings—
And why the sea is boiling hot—
And whether pigs have wings.'

Lewis Carroll
Through the Looking-Glass

Alice and the Quantum Cat brings together many of the leading interpreters of the New Physics with the goal to make the miraculous mysteries of the invisible looking-glass inner world *tangible* for the non-scientist.

The project got started when Dianne Collins and I read an interview with Fred Alan Wolf in the *Yoga Journal* in 1989. Fred described how the act of observation collapsed the wave function from infinite possibilities to an actuality. This fact unleashed a passionate curiosity and mission—to bring these ideas into the mainstream, so in 1993, we invited Fred to spend time with us at my house in Connecticut to birth a quantum Alice.

In the years to come, as I learned more about this astonishing new world I 'lost my grip on reality' and began to experience the universe as an infinite reservoir of living energy, potential and creativity. While that transformation was not painless, what I gained from the wonderful author-scientist is beyond measure.

The product was *Lewis Carroll's Lost Quantum Diaries*, published in German in 1999 under the title, *Alice Zwischen den Welten*, by DVA, and in Japanese in 2002 by Tokuma-Shoten. And now, finally, after her blue-plate makeover led by master surgeon F. David Peat, Alice is at long last coming home to Anglo-Saxons and their compatriots on both sides of the pond.

So, lads and ladies, we invite you to set aside the concerns of 'real life' for a time, close your eyes, focus your intent, and birth a star in the sea of infinite possibility.

---

6. Ibid, p. 22.

Oh, and lest we forget, Mr. Carroll, it's time to break out the champagne, lift a glass, wish our brave voyagers well, and toast your inspiring, ever-sweet genius!

New Haven, Connecticut
Thanksgiving, 2010

PREFACE

F. David Peat

Alice's Adventures in Wonderland was first published in 1865. It was written by the Reverend Charles Lutwidge Dodgson, a talented mathematician and logician with a lectureship at Christ Church College, Oxford. The story was originally handwritten as a present for Alice Liddell, the daughter of one of Dodgson's Oxford colleagues but later revised and expanded for publication under the pseudonym of Lewis Carroll.

Dodgson's book was filled with a variety of logical jokes and extravaganzas as well as parodies of Victorian poems and a little fun at the expense of some of his friends and colleagues. Probably the most interesting version of the book is that annotated by Martin Gardner who explores the various mathematical and logical in-jokes that are scattered throughout the text.

It was this mixture of fun and erudition that prompted William Brandon Shanley to invite some leading scientists and science writers to speculate on what Dodgson may have done with his story had he been aware of the development of quantum theory, and later chaos theory, in the twentieth century. But how to account for all these latter-day Alices, each speaking in a rather different way and located in everywhere from San Francisco to a railway carriage in Ireland, as well as in different time periods? Dodgson's own book holds the clue with its Cheshire Cat that can vanish while leaving only its smile behind. Could this be the very same cat that Schrödinger proposed putting in a box in order to illustrate a fundamental paradox of quantum theory via a thought experiment?

Schrödinger's box contains a radioactive substance whose decay will release cyanide into the box. While we, the observers, do not know if the cyanide has been released or not, quantum theory dictates that the state of the cat in the unopened box must include all possible combinations of being alive or dead. Technically this is known as 'the superposition of the wave function.'

Yet when we open the box we see only a live cat or a dead cat. This is the famous 'collapse of the wave function'—the emergence of a unique solution from a superposition of all possibilities. How can the paradox be resolved? A number of solutions have been proposed including one thought up by Hugh Everett and John Wheeler in which the universe is constantly splitting. In some of these universes there will be a live cat and the human observer, in others a dead cat and the human observer.

And there you have it: with the help of the Cheshire Cat Alice is able to visit several of those multiple universes in each of which there is a slightly different Alice. So welcome to the story of Alice in her multiple worlds.

Pari, Italy
2010

PART I:
THE QUANTUM UNIVERSE

1 BESIDE THE RIVER BANK

F. David Peat

In which Alice meets the Quantum Cat.

Alice was sitting on the river bank listening to the sound of the water and turning the pages of her book. It was a book with pictures in it, which pleased Alice but there were far too many foreign names in the story, names that Alice found difficult to pronounce such as Wittgenstein and Schopenhauer. What on earth have they to do with an adventure of a little girl who falls down a rabbit hole, thought Alice?

In the end Alice set down her book and decided that it was just the right time of the afternoon for a little doze. No, not a doze, really more a time to rest her eyes. Indeed she was about to close them when suddenly she saw a smile—just a smile but then a moment later some whiskers and then two eyes, and suddenly there was a cat sitting beside her on the grass.

'Am I in that book?' asked the cat.

'But you can talk,' said Alice, 'and by the way, I don't think it's very polite to sneak up on people like that.'

'Well, am I?' the cat asked again.

'I don't think so,' Alice replied, 'on the other hand I haven't finished it yet. But there is a rather horrid pessimistic little man, a German philosopher, named Schopenhauer.'

'Oh yes, I know him. He wrote about me, but he wasn't really that bad. Not like that dreadful Erwin Schrödinger who kept me locked up in a box.'

Alice leaped to her feet. 'So you *do* know the book. Schrödinger's in it, too. But why did he put you in a box? That's not a nice thing to do.'

'Sit down again and I'll tell you the whole story,' said the cat, 'but you mustn't interrupt. Well, to begin with, Schopenhauer. He was very kind to me and wrote that I was the same cat who also used to play in his house two hundred years earlier.'

'But that's not possible,' cried Alice, 'cats can't live that long, they're not elephants or tortoises.'

'I thought I told you not to interrupt,' said the cat quite crossly, 'I'm no ordinary cat since I can be in two different times at once, and in two different places for that matter. And what's more when I catch mice I make quantum leaps. And that's where Schrödinger came in. He saw me doing a leap one day and shouted "quantum cat," then put me in a box with a very nasty bottle of cyanide. If the bottle broke then I was supposed to die, and if it didn't I'd live to see another day. But Schrödinger's box was a very special box, it was a quantum box.'

'That's the third time you've used that word—what is a quantum?'

'Oh, it hasn't been invented yet, well not in your time. But if you'll stop interrupting I'll finish the story. You see in the quantum box I could be both alive and dead at one and the same time. And even 99% alive and 1% dead, and even...'

'But that's...'

'Stop it,' said the cat, 'I know what you're going to say. What did that nasty Schrödinger see when he opened the box? Well, I told you I could be in two places at once didn't I? But it's even better than that because I can exist in a whole series of parallel universes at one and the same time. In some I'm alive and in some...well, we won't go into that now. No, better still we will. I'm going to show you. Now let me sit in your lap, there. Blink your eyes and...'

Alice blinked once and suddenly felt herself spinning and falling, falling towards a bright light, then...

2 ALICE MEETS PROFESSOR FLOW...

Fred Alan Wolf, William Brandon Shanley and Dianne Collins

...and receives her first lesson in how Quantum Possibility turns into Everyday Reality.

Alice rubbed her eyes, trying to adjust to the light. She could just begin to make out an image of... 'Why, yes! It's a man,' she determined. 'But that's odd. He's running up a set of stairs. But wait! He's running up a set of *down* stairs!'

'I can't believe that!' Alice cried out, thinking she'd lost her mind.

'To get around here in this Quantum Universe you might as well lose your mind,' he called out to her, 'or your grip on reality!' he laughed.

As his descent slowed, she was able to examine the mysterious man more closely. He was bearded and bespectacled and appeared to be about fifty, wearing a white lab coat. Its pockets were filled with watches, calipers, rulers, micrometers, and all kinds of other measuring devices. He was clutching a camera in his hand. Several others were hanging from his neck.

'My grip on reality? Whatever do you mean?' Alice shot back.

'What do I mean? Because the chances are that right now, everything you know about the universe and its laws, and how they relate to matter, mind and consciousness, is more than 99.99 percent likely to be completely wrong! So, by all means, *please* lose your mind, but make sure to hang on to your intent and your imagination!'

'But wait!' The quirky looking man froze in his tracks on his futile trek. 'What is this place you call the "Quantum Universe" that I've somehow managed to get myself into?' asked Alice.

'The Quantum Universe is the magical and paradoxical subatomic domain of the quantum where parallel lines converge and things relocate without traversing space. Here, measurements like time/space, energy/matter are only quantifiable in very peculiar ways. Phenomena begin to phase out of the everyday classical, Newtonian reality you live in, and become invisible to you.'

Alice's curiosity was piqued. 'Quantum Universe, you say? What does "quantum" *mean*, anyway?'

'Quantum is the Latin word for "how much." In Quantum Theory a quantum is the smallest individual unit of energy that can be associated with any single subatomic event,' the unusual man informed her.

'Well, that's not very helpful,' Alice decided, thinking he'd used lots of words to communicate very little. 'How much of what? And just what is Quantum Theory?'

Deciding to take time to enlighten the young lady more fully, the bespectacled man floated through space for a more intimate elucidation. 'Quantum Theory is our modern theory of matter and light. We know the universe is made of matter and energy, and that matter and energy move in quanta, or bursts. An electron is a quantum of electricity, for example. A photon is a quantum of light. Quanta move in the universe, not in a continuous movement, but in tiny, explosive, discontinuous leaps without traveling in between. A quantum is actually *both* a particle, small and determined, *and* a wave, spread out in space-time.'

Alice thought for a moment. 'How could something like a particle move from one place to another without traveling in between and be both in a fixed place and yet spread out in space?'

'Quantum systems are spread out all over space and time,' he answered, reading her mind. 'A photon is both a wave and a particle, a particle both here and there, now and then.'

'But that's not possible,' she said.

'But it is true. And that's just the beginning. Mind plays an important role in physical observation. There's a creative role between the observer and the observed. When a quantum physicist sets up an experiment, he decides beforehand what kind of experiment it will be. If he does a wave experiment, he gets a wave answer; from a particle experiment he gets a particle answer. The type of observation he uses evokes one or other of the underlying possibilities. This is called the "double-slit experiment." You see a quantum system, like a photon, actually follows all possible paths, all at the same time. It behaves as if it was smeared out over space and time and is everywhere at once.'

Alice was flummoxed by the zany man's claims, but remained curious.

'People used to think that an atom consists of a nucleus of protons and neutrons circled by electrons in fixed orbits. Today we know that the thought picture of classical physics was mistaken. In the Quantum Universe, the atom is an amazingly complicated, finely balanced assemblage of forces and particles performing an exquisite ballet woven in four-dimensional space-time: particles of light are "photons"; "phonons" are quantized sound waves; other elementary particles include protons, neutrons, muons, pions, leptons—the list is likely endless. For it seems the more deeply we plumb the depths of Mother Nature's inner world, the more particles we find. In our new quantum thought picture, the inner universe of elementary particles is alive, creative, interconnected, interpenetrated, and communicating with itself at every level of existence faster-than-light.'

Now Alice was really confused. 'I learned in school that everything is made of different parts...'

'...of dead matter that somehow came together and gave birth to life,' he said, completing her thought. 'Is that right?'

'Well, yes. At least that's what I think I remember.'

'Alice, all matter and energy in the universe are instantly connected and in a constant creative dialogue with their environment. Moreover, energy seeks to fulfill its potentials in all possible variations.'

'But you're speaking about these little particles as if they had minds all their own,' Alice chided.

'Let me give you some background to help you understand what we now know for certain about these basic constituents of matter and energy in the universe. May, I?'

'Yes please,' she said eagerly.

'When you looked around you saw a world of solid objects, which are actually dynamic energy patterns dancing in universes of what you thought to be "empty space." Now imagine that an atom, the size of a football stadium with an electron the size of a dust mote spinning on its rim. The nucleus would be the size of a golf ball on the fifty-yard line. Everything in between, according to our senses, appears to be "empty space." Yet, according to the maverick quantum physicist David Bohm, one cubic centimeter of this so-called "empty space" contains the potential to produce more energy than we see in the entire visible universe! Such is the infinite potential and miraculous nature of the Quantum Universe!'

Alice was dazzled by the implications. 'That's just incredible!'

'Our modern picture of light and matter, Quantum Theory, describes reality as possibility waves when not looked at, as actual particles, when looked at. Behold these four fundamental Quantum Facts:

A quantum object, an electron for example, can be in more than one place at the same time. That's called its *wave property*.

A quantum object cannot be said to manifest in ordinary space-time reality until we observe it as a particle. That's called the *collapse of the wave function*.

A quantum object ceases to exist here and simultaneously appears in existence over there; we cannot say it went through the intervening space. That's called the *quantum jump*.

And, a manifestation of one quantum object, caused by our observation, simultaneously influences its correlated twin object—no matter how far apart they may be. That's called action-at-a-distance, or *quantum non-locality*.

Here, in Quantum Reality, atoms are less like particles and more like promises that have only possible positions until you observe them,' he informed her. 'Moreover your ordinary reality is totally dependent upon these promise-like qualities of Quantumstuff throughout the entire universe.'

'I can't believe *that!*' Alice found herself repeating. Suddenly Alice was standing beside the White Queen in Wonderland!

'Can't you?' the Queen said in a pitying tone. 'Try again. Draw a long breath, and shut your eyes.'

Alice laughed. 'There's no use trying,' she said. 'One *can't* believe impossible things.'

'I dare say you haven't had much practice,' said the Queen. 'When I was your age, I always did it for half-an-hour a day. Why, sometimes I've believed as many as six impossible things before breakfast.'

'Impossible!' Alice heard herself say. In the next instant she was back with the man who was saying such confusing things.

'I was taught that if I can't see something, feel it, touch it, or taste it—it's not real. If I can't kick it—it's not there. What you say is sheer nonsense.'

'Pushing reason to the point of absurdity, maybe, but not nonsense.'

Alice looked around for something real, and settled on herself as the only thing she could count on. 'I know I'm real,' pointing to herself, 'and you—you're real. Wait a minute,' Alice stared hard at the man, and pointedly asked, 'Who are you, anyway?'

'Ah, let's see,' the curious man said, pondering for a moment. 'I'm a trillion, trillion, trillion molecules each adored by at least a dozen electrons. I'm a live-wired net of ten billion neurons telling a million billion body cells what to do—while enjoying a multi-cultural, multimedia live-action experience, part of which is the peculiar conviction that I am me. Just like you!' he chuckled. With a glimmer in his eye, he lovingly held his gleaming calipers up high, as if to measure infinity.

'My name is Professor Fred Alan Flow, metaphysicist extraordinaire!' he boasted. Obviously pleased with himself, he turned his attention back to Alice. 'And your name is Alice...'

'Yes, but how did you know? And I know I'm more than just a bunch of thoughtful electrons with possible positions spinning around in space! My name is Alice B. Alice and I live on Classical Drive in Newtonville. But wait! I can't be that Alice—because I'm here!'

'We're in Quantumland now, not Newtonville, Alice,' counseled Flow.

'And just what do you do here in Quantumland, Professor, besides confuse young ladies?'

'I explore the paradoxes and intricacies of Quantum stuff in a myriad of mysterious mindscapes,' he gestured around. 'Whether it be the universes of Quantumland, Chaosland, Biologyland, Forgotten Truth, or the Other Side—to name just a few—and now of you, I'm searching for answers to the mysteries of Creation in the mirrors of the Mind.'

'This Quantumstuff sounds so confusing and unnecessary!' Alice exclaimed. 'I want to go back to the way things were before, separate and predictable, not all mushed up in some nutty physicist's madhouse!' She thought for a moment. 'And why bother with it all. I'll never use it in real life anyway.'

'Why?' Flow asked. 'It's very important, Alice! Everything in the physical universe must move in the fashion I've described or cease to exist! Since you and I are composed of atoms and subatomic particles, we too, must "take the quantum leap." Like the one you took to get here.'

'I did?' Alice wondered, then she remembered what the world was like before she popped into the Quantum Universe. 'Well, I suppose you could say my state did change. But why is this so important to you, Professor Flow?'

'Because I'm working on a riddle beyond space and time, energy and matter, and number. It's a real puzzle of a mind-bender distinguishing and defining a domain where someone's conscious intent collapses pure possibility into Reality.' Glancing at her sideways, he added with a knowing air, 'I should think you would find all this very pertinent, my dear.'

Before Alice could question him further, Flow held up his camera.

'You see, when the Wiff! pops, Reality happens! It's like magic! The Wiff! is the invisible, spread-out vibrating entity of transition present just before a wave of possibility collapses into an actual particle. If I can just catch a picture of the elusive joker, I think I can solve the riddle of the universe, define consciousness, discover the Grand Unified Field, end the search of science, and put my flag on the mountaintop.'

Suddenly, a flash of blinding light swept through from nowhere and everywhere. 'What was that?' gasped Alice.

'Young lady, you've just spotted the Wiff!'

Instantly, a flash of light beamed between them with a loud ZZ-ZAAAPPP! Out of the blinding light came a multiplicity of breathtaking images projected into space. The Wiff! was be-bopping and scat-singing in riddles and rhymes.

'I'm the Wiff!—the possibility of all probabilities, possibly! Or you might say, I'm the probability of all possibilities, probably! Or how about, possibly, the possibility of all possibilities?!!'

Next, the Wiff! changed from a featureless glow into a dazzling display of holographic imagery: a dancing pink elephant, a marching band, and a cheering crowd.

'I'm magical, mysterious, mystical, and marvelous! I'm the mathematical function that turns possibility into reality—the potential for anything to manifest,' said the Wiff! 'In less than a twinkle of the eye, I can be this, or I can be that. When nobody's watching, I'm formless. Then when you look, I take form! Close your eyes and hold on...'

In the next instant, Alice and Professor Flow were flying high above the west coast of the United States. A long wave front stretched from Alaska to Mexico. 'Now all my energy is spread out across thousands of miles of coastline,' said the Wiff! 'Now watch me make magic happen when you look!'

When Alice and Professor Flow opened their eyes, all the energy of the thousands of miles of wave front collapsed into a single grain of sand at Radio Beach in Santa Cruz, making it jump hundreds of feet into the air.

Alice heard Flow chuckle. 'It was bizarre thought pictures like this that bothered folks back in 1905 when Einstein explained the photoelectric effect with his concept of light quanta. How does all the wave's energy get so concentrated so suddenly? Only the Wiff! knows!'

The Wiff! vanished as quickly as it had appeared. Professor Flow noticed that Alice's curiosity was aroused by the Wiff's magical display, and seized the opportunity to explain more of the fundamentals of quantum reality to her.

'In quantum reality, Alice, a moving particle like an electron doesn't travel from point A to point B to point C and back again, but exists—or doesn't—at all three points, and all points in between, at any given moment. It has the literal ability to be in more than one place at a time. It can also exist at all available energy levels and in every direction of spin at once.'

Alice was astonished by what the professor was saying, so Flow continued: 'So, at this level of reality, Alice, matter exists only as a possibility, a possibility that somehow seems to fill space, until we become involved.'

What do you, or I, or anyone for that matter have to do with the behavior of matter, anyway?'

'As the Wiff! so aptly demonstrated, matter becomes more "real" when we become involved, Alice. The act of observation doesn't merely reveal a particle's condition, but actually determines it, forcing it to select just one of its possible states.'

'But I still don't understand how an act of observation by me could determine the condition of matter or energy,' Alice offered skeptically.

'Early in the twentieth century, Alice, quantum physicists such as Werner Heisenberg discovered that atoms are not "things." It became clear that an atom is partly a construct of thought and partly a kind of material substance. So the prime question became: how do thought and substance interrelate? It led Heisenberg into postulating the idea of a middle realm, which I now call "the imaginal realm," where our thoughts about the way things are begin to relate to each other in a certain sense. It would appear that thought is a mark of consciousness itself. So it would appear that matter is some denser form of consciousness itself. And I think this is where the idea begins to get more exciting.'

'But I *still* don't understand how matter can become more *real* when we become involved? Show me what you mean, Professor Flow.'

'Let's look for the electron that's making the pattern in the atom over there.'

Alice looked over at a packet of vibrating energy next to them. 'How do I do that?'

'Just ask the electron.'

With that, Alice and Flow were inside the atom.

'Electron?' asked Alice, 'we're inside a cloud of pulsating light.'

'That's right, Alice.'

'Now ask the electron where it is.'

'All I see is this strange cloud all around us, a luminous yellow fog, like a will o' the wisp! Where are you, electron?'

'Could you be more specific,' the electron asked, changing color and intensity in rhythm with its voice.

'Are you standing still at the moment?' asked Alice.

'Well, nooooo, not reeaaallllyyyy.'

'Well, is it moving then?' Alice asked turning to Flow.

'Ask it.'

'So, you're moving around then?'

'Well, no. I'm not actually moving around either.'

'But surely you have to be either moving around or standing still. You have to be somewhere, don't you? It doesn't make sense.'

'Well, I'm probably somewhere,' answered the electron. 'And I'm more likely to be there than somewhere else. But I could be somewhere else, too. In fact, I'm lots of places at the same time, with varying degrees of possibility. And in another sense, I'm everywhere.'

'No come on, where are you exactly?' asked Alice.

Professor Flow turned to the cloud of matter and stared at the center intently. In that instant, the cloud disappeared and an electron, a point of light, popped into place, and disappeared. Then the cloud appeared again, spreading out from where the electron was.

'Oh, I see, Professor Flow looked you into existence.'

'Right! At the moment he looked, that is where I was,' the electron informed Alice.

'OK. But how fast were you moving when you passed that point?'

'Ah, well, sorry. I can't show you that. But if you don't try to see where I am, I can show you how fast I was going. And vice-versa.'

'Stop professor! Now I think you're both losing your grip on reality!' *And so am I,* she said silently to herself.

'That's right, Alice, quantum physicists have lost their grip on reality, and now finally you're doing so, too! Because the basic building blocks of matter, the stuff of rocks and air alike, are nothing more than what physicists have dubbed possibility waves, conceptual cloud-banks that don't exist in any *real* sense until they are observed by a conscious being, Alice. Now, would you like to see one of these atoms even closer?'

'Why yes, I would.'

Flow stretched out his arm and plucked a fuzzy ball out of the air. He brought it close to Alice for her to look at. 'Look closely, and you'll see how you can change it.'

All Alice could see was a blurry cloud of glowing light. 'But I don't see anything. It's just a blur. I'm confused again! What am I looking at?'

'Just a moment, Alice. Now I want you to imagine that you are looking at the energy of this atom in my hand.'

'I know we've talked about it, but what is energy, anyway?'

'Energy is like what you feel in your body.'

'You mean I should look at this ball as if I'm feeling it in my body at the same time?'

'Yes. What you are feeling right now?'

'Well, I'm feeling a little tired...'

'Now look at the atom, Alice.' The atom suddenly transformed from a confused dance into a pulsating, glowing sphere.

'Oh my! Did I do that?'

'Yes,' Flow and added proudly, 'just by your observation.'

'That's exciting!'

'Now that you're excited, look again.' The atom grew in size, and looked like a spinning dancer with a tutu spinning around it.

'What happened?' asked Alice.

Professor Flow was explaining, and she pushed her doubts aside to listen. 'When you got into an excited energy state, the atom got into an excited energy state.'

'I know seeing is believing, or is it believing is seeing?' Alice was confused again, and ever dubious. 'I still don't believe you. That wasn't me! You're playing tricks on me. That was you. You're a magician, aren't you?'

'No, I'm a physicist, Alice. And the kind of miracles at the very heart of creation remains a mystery to us. But there are some things we do know. Things are mere possibilities until they are observed. The form the basic energy units of the universe take seems to depend on what they are asked to do. A quantum system's potentials, or "propensities" all coexist simultaneously and all play a part in the evolving dynamics of the systems until some act of measurement selects one of them to be the next actuality.'

'But what does all this confusing quantum physics stuff *mean*, Professor?'

'Quantum physics tells us that we are all connected to one another, and to everything in the universe. That consciousness is not something separate, but an intrinsic element of all matter. That in some way we don't yet understand, we all touch each other, no matter how far apart we may be. Our efforts not only lead to the building of skyscrapers and rocket ships, but our very thoughts reach out and touch the universe. But wait...'

In a panic, Professor Flow looked at one of the watches. 'Oh, dear! Oh, dear! I shall be late!' as he turned to hurry on.

'But wait, Professor,' Alice implored, not wanting to be abandoned in this world of Wiffs! and schizophrenic quantum bits. 'You look like you know where you're going. Would you tell me, please, which way I ought to go from here?'

Flow ceased his impossible trek running up the down stairs for a moment. 'That depends a good deal on where you want to get to,' he responded sagely. 'And who do you want to be when you get there?'

Alice started. 'Why, *me*, of course. Who else would I be?'

'Hm. Good question. Who else would you *like* to be, my girl? Here in Quantumland you will learn new ways of seeing that may open up new possibilities and new ways of being that will get you back from whence you came, but I remind you, you will never be the same again. Shall we get started?'

Alice, still confused caught up in the possibilities, simply nodded her head, starry-eyed.

'We gotta go and never stop going till we get there,' said Flow.

'Where are we going, doctor and what sort of people live here?' asked Alice.

'Well, Alice, you're about to find out the Quantum Universe is full of strange characters with outlandish quantum stories to tell and astonishing experiences to share. But I warn you, getting any two quantum physicists to agree on a description of Quantum Reality is like "an impossible voyage of an improbable crew to fund an inconceivable creature!" They'll tempt you with their thoughts and you may even meet some who'll dispute the experiments I've shown you. "Balderdash and bandersnatch," they'll say. As they'll Snark-fit all day.'

'But I just saw them with my own eyes,' Alice objected. 'Now you're telling me it was only a magician's trick, or hallucination, after all?!'

'Physicists themselves cannot agree on what is real and what is nonsense down here at the quantum level, although they all agree about the theory on the human scale and the four Quantum Facts—the quantum wave property, the collapse of the wave function, the quantum jump, and quantum non-locality. Other than those four, in Quantumland, as physicist John Wheeler said, "There is no law except the law that there is no law!" This place is wild! Indeterminacy is built into Quantum Reality; it's an inherent feature here. The universe is not only stranger than we imagine, it's stranger than we *can* imagine.'

'Stranger than we *can* imagine?' Alice asked, 'Just how do you propose I find my way, then?'

'Physicists have dreamed up and devised at least eight different tentative theoretical pictures, or maps, of the quantum world, Alice. And even though many have important elements in common, they are each profoundly different and all quite bizarre. So, keep your wits about you when you open new doors and enter these strange new thoughtforms. Your own way, like the Quantum Universe itself, remains a mystery; an enigmatic riddle hidden inside a puzzle wrapped in a conundrum. Explore these worlds, ask your own questions, and let your inner feelings guide you. But don't worry, Alice, you'll be OK! Instead of the "nightmare of determinism" you'll be dreaming about possibilities.'

Alice thought for a moment. Even though she was afraid and couldn't quite grasp nor fully understand everything the wizardly Flow was saying, she knew she trusted her own intuition and feelings. 'Ask my own questions. Let my intuition guide me. Discover my own path. I'll be OK? Is that right Professor Flow?' Alice asked.

'By George, I think you've got it!' Flow twinkled. And with that, he opened a door with a sign that read: Wormholes, Worldtubes and Other Portals to Possibility.

'After you, my dear,' invited Flow, indicating a glowing energy tunnel, pulsating before them.

3 DEEP REALITY RESEARCH

Nick Herbert

In which Professor Who explains how Observation turns Possibility into Actuality.

When Alice stepped through the door into the energy tunnel at Professor Flow's bidding, she found herself in a long corridor. When she looked behind, the door was gone, and so was Professor Flow! Alarmed at first, she realized there was nothing else to do but continue down the hall until she came to a door marked DEEP REALITY RESEARCH. Below was a colorful sign, which read: VISITORS WELCOME and below that an official warning: AUTHORIZED PERSONNEL ONLY.

She pushed open the door and entered an anteroom hung with pictures of scientific instruments, blueprints of fantastic devices and photos of Indian gurus in exotic settings. She opened the second door and entered a laboratory. The laboratory was full of the usual blinking lights and strange bubbling potions interconnected with a myriad of flexible tubes that resembled a nest of caterpillars (or hookah hoses). A thin graying man with bright blue eyes looking something like a very alert toad was seated in lotus position in front of a computer. He whirled around to greet her.

'Who are you?' asked Alice.

'I'm Professor Who. Who are you?' he responded, reminding Alice of an owl.

'I'm...I'm...Who did you say you were?'

'Who. Who. Professor Who. And you? You? You are who?'

'Oh, your name is Who? Like the science fiction Dr. Who?'

'Exactly, yes. Except that Dr. Who is fictitious, but I—Professor Who—I am real. I am the *real* Who. And who is the *real* you?'

'My name is Alice, Professor. I saw the sign on your door, 'Deep Reality Research' and...'

'Are you interested in reality, Alex? Not many people are interested in reality these days. It is fortune; it is fame; it is this thing; it is that. Everyone is chasing something but very few are looking for the Real. If you are seeking reality, Aloysius, you've come to the right place.'

Realizing the professor could not remember her name, Alice giggled.

'Reality is what we study here. Not things, not people, not theories about things, not theories about people, but Reality itself—what makes it all work.'

'What got you interested in reality, Professor?" Alice was curious.

'Oh, Albert, that is a good story!' he exclaimed, clapping his hands with glee. 'Did you know that I was a student of Werner Heisenberg?'

'Well, no,' Alice admitted.

'From Heisenberg himself I learned quantum physics. I was one of his best students. I might have been as famous as he except for my interest in philosophy.'

'Philosophy?' she asked astonished.

'Yes, philosophy, Alicia. I was also friends with Moritz and Max in the philosophy department. Max and Moritz were doing experiments with mescaline sulfate, one of the so-called hallucinogenic drugs. Yes, with Max and Moritz I went for an outing in the Black Forest, an outing from which I have not yet returned. That was in 1925. For me, the Summer of Love. Nineteen twenty-five was the Summer of Love for physicists too, for that was the year that Heisenberg discovered the Great Secret, the mathematical structure of the quantum world. Heisenberg called it 'Matrix Mechanics.' Physicists now for the first time could actually calculate and predict the behavior of electrons, protons and photons of light, the modern theory of the entire material world.'

'So instead of working with Heisenberg, you were doing drugs in the woods, Professor?' said Alice, allowing a touch of scorn to creep into her voice.

'It was the Summer of Love for physics, Alonzo,' he said by way of explanation. 'Then a year later Herr Schrödinger discovered Wave Mechanics. Another theory of matter and light very different from Heisenberg's, but making the same predictions. Physicists were utterly confused about the quantum world for a quarter of a century, then two complete theories of the world in two years! Nothing like that had ever happened before in science. We later discovered that these two theories were just two different ways of looking at the same core truth called Quantum Mechanics. Do you understand Quantum Mechanics, Alberta?'

'Quantum Mechanics? Why, no, I can't say that I do,' she confessed.

'Here, let me show you a video.' Professor Who picked up a remote control and an animated film began to roll on one wall.

The announcer on the video began his narration in an avuncular tone: 'The old-fashioned Newtonian world was built of particles moving in fields of force such as gravity, magnetism and electricity.'

On the screen, multi-colored billiard balls moved haphazardly. A man holding a magnet swerved some of the particles. A woman rubbed a balloon on her skirt and swerved some of the particles. A picture of particles radiating 'energy waves' appeared on-screen, and the narration continued: 'From each particle a pattern of force fields emanates. And each particle's motion is completely determined by the fields it feels from all the others. Completely determined means completely predictable. Once you start it up, this world of mutually interacting particles and fields behaves like a giant machine, like a game of billiards. The world according to Newton behaves like a giant clock—perfectly picture-able…'

Animated particles and fields gamboled on the screen.

'…and perfectly predictable, forever and ever. And most important, this world, this clockwork, behaves exactly the same whether you look at it or not—it's always made of particles and fields. In Newton's world, "looking: is nothing special—just one interaction among many."'

'This is important,' interrupted Professor Who. 'Observation doesn't change the world of Newton. That's very different from…'

'Shhhushhh! I want to listen,' Alice interjected irritably.

As if on cue, the narrator continued: 'The quantum revolution began when the clear distinction between particle and field was seen to break down. Max Planck and Albert Einstein showed around 1900 that light—which had been measured for centuries to be a wave—a wiggle in the electromagnetic field—actually behaved in some experiments like a particle called the "photon."'

On-screen waves were going behind a blue curtain, emerging as particles, then going behind a red curtain and emerging again as waves.

'Then the recently discovered electron—which was certainly a particle, its mass and charge had been accurately measured in England in 1897—was shown at Bell Labs in America to behave like a wave: fast electrons have short wavelengths; slow electrons have long wavelengths.'

The screen showed two beams of particles—fast electrons and slow electrons—going behind a red curtain, then emerging as short and long wavelength waves. They continued on behind the blue curtain to become particles again.

'By 1926,' stated the announcer, 'the biggest part of the puzzle had been solved: Heisenberg and Schrödinger independently and in different ways discovered the basic mathematical description of the particle/wave. But it took several more years to make sense of the mathematics. That process is still not over. We still do not know exactly how the quantum world works. Here is our best guess. Whenever a quantum object is not being looked at, it behaves like a wave. This wave is not made of real substance but is more like a catalog of possibilities. The ebb and flow of these invisible quantum possibilities is described by Schrödinger's wave equation or by Heisenberg's matrix mechanics.'

On the screen was a complex colored wave animation: a shimmering fractal imagery. It repeated and repeated itself in ever smaller patterns until it was like looking into a world of infinite complexity.

'When a quantum object is looked at, it always appears as a particle—a particle that is more likely to appear where the wave is large and less likely to appear where the wave is small. The position where the particle actualizes seems to be utterly random. This so-called "quantum jump" (wave to particle) seems to be governed by chance: here, in Einstein's words, is where God plays dice with the universe.'

'But I thought the Wiff!...' Alice began to interrupt, but caught herself.

A series of waves broke against the screen. Each one left a single particle behind. Soon, a wave-like diffraction pattern became clear, made of patterns of particles, building up on the screen.

'You see when the waves meet they do a sort of dance, peaks meet troughs and cancel out, peaks meet peaks and get even bigger and...'

'I think I understand,' said Alice, 'it's a sort of dance of waves that build up those beautiful patterns.'

'But in the quantum world, observation is crucial, Alice. Observation changes the universe. Before you look, the universe is not even real; just a shimmering pattern of vibrating possibilities. Then when you look, it freezes into a static pattern of particles, only to return to vibrating possibilities when you avert your gaze.'

Alice saw an animation of the observer freezing fractal waves into dots, and then dots flowing into fractal waves when the observer looked away. She was mesmerized. The announcer droned on.

'When you don't look, the quantum world consists of waves—not real waves, but waves of possibilities. When you do look, one set of these possibilities is realized—at random—in the form of particles. This new view of the way things work is called quantum physics—the most successful theory of matter ever invented.'

As Alice watched, a diverging fractal wave poured towards her, transforming into a pattern of particles, which slowly faded to black.

'So the world is possibility waves one moment, and actual particles the next?' asked Alice.

'Yes Annabelle. It turns into particles when you look,' nodded Who sagely.

'Well what kind of a world is that?'

'It's called "Heisenberg's Duplex World."'

In that instant, the scene peeled away and a slight man in a suit stepped into it.

'Well, hello, Werner!' exclaimed Professor Who. 'How am I doing? Did I get it right?' The man spoke with a German accent: 'Just fine, don't worry about it,' Heisenberg said with the wave of his hand, 'but I thought Alice might prefer to hear about my quantum model from me directly.'

'Why yes, I would, and very nice to meet you,' Alice said with a curtsy.

'Nice to meet you as well, Alice. You see, I realized that the mathematics was right but that language is a serious problem, because atoms cannot be described in ordinary language as "things." Atoms and elementary particles are not as *real* as the phenomenal world; they form a world of potentialities and possibilities rather than one of "things" and "facts." The probability wave means a tendency for something. My quantum model is a quantitative version of the old concept of *potentia* in Aristotle's philosophy. It introduces a strange kind of physical reality just in the middle between possibility and reality.

In a flash of light, Werner Heisenberg was gone. 'Was that really Dr. Heisenberg?' Alice wanted to know.

'Why yes, it was, and I'm amazed that he would honor us with his presence. But let's get back to our conversation,' said Who.

'But Professor Who, I thought it was the Wiff! that turned waves of possibility into actual particles when observed…'

'The Wiff! Oh, No! You must have been visiting with that nutball Flow!' Who surmised.

'Why, yes I have, and he was really quite nice. But he warned me about quantum physicists not always…'

'The Wiff! What preposterous nonsense. A figment of Flow's overactive imagination. Why, to my mind, he's a complete sociopath! Forget about the Wiff!'

Alice left that one alone, and went back to her exploration of quantum reality.

'But if we can never see these waves of possibility, how do you know that the un-looked-at world is made of waves?' she challenged.

'It's the mathematics. If you describe the un-looked-at world as waves you can predict what it will look like when next you look.'

'Really?' she was doubtful.

'Yes. Also, I have discovered a way to actually see quantum waves.' He pumped up proudly.

'How can you do that? How can you look without looking?'

'In the Black Forest, I saw directly into the nature of Reality. I was actually able to see the world as it is—as shimmering waves of possibilities.'

'You mean on drugs?' Alice asked flatly.

'Not only on drugs, Acacia,' he replied defensively. 'I traveled to India and studied with many wise souls. Now I can see the vibratory nature of reality in states of deep meditation.'

'Should I call you "Who the Guru" then, Professor?' Alice teased. 'Inside your head you can make up anything and believe it to be true! How do you know you're not just fooling yourself? Maybe these vibratory possibilities you saw in the Black Forest were just hallucinations.'

'My visions carry such a sense of certainty,' he stated firmly, 'their truth is difficult to deny. But as a scientist, I must agree with you: I may be fooling myself. That is what these machines are about. I am devoting my life to discovering a way to directly probe the quantum world, to penetrate to the realm of real reality, to investigate first-hand the inner world of vibratory possibilities.'

'You mean you want to somehow look at the world without really "looking?" Is that it, Professor Who?'

'Quite right, Allen. To look without looking. To sneak an unnoticed peek behind the veil of Maya.'

He acts for all the world like a naughty boy trying to look up a lady's skirt, thought Alice. Out loud, she politely asked: 'How close are you to succeeding, Professor?'

'Alas, Azuza, all my experiments have failed. No matter how carefully I try to look, all I ever see is particles. However, I could teach you some meditation techniques so you could see the quantum waves for yourself. Just sit down on this pillow, close your eyes, and repeat after me.'

'No thank you, Professor. I don't have time for that. I must be off,' Alice said quickly, scampering away from the eager Professor Who.

'Thank you for your fine company, Allegra. It sometimes gets lonely down here. I once had a dog named Allegra,' he muttered. 'Or was it a cat?'

At the word cat Alice suddenly thought that she should remember something. Then it came to her and she blinked.

4 OYSTER QUADRILLE

Nick Herbert

In which Alice interrupts an argument between two conflicting Realities: One created by Mousetraps, the Other created by Minds.

Alice blinked and found herself on the river bank again with the Cheshire Cat on her lap.

'I'm not at all sure if I like those parallel worlds at all. In a way it was me but at the same time it wasn't me,' Alice admitted.

'I know exactly how you feel,' said the cat. 'I've just been to Tiffany's with my friend Holly Golightly. I often go there but it makes me very hungry. You don't have a mouse in your pocket by any chance?'

'Hungry, that's right. I feel that I could eat a horse—well, a very small horse. But tell me a little more, dear cat. You explained all about splitting off into different universes but I think you also told me you could be in different times as well when you were with Mr. Schopenhauer.'

'Oh yes. I was both the cat playing in front of old Schopenhauer and the one that played on the mat two hundred years earlier.'

'And could I do that?' asked Alice.

'But you will, you did.' replied the cat.

'But that means I'd get the best marks in school because I could jump ahead and read all the exam questions and I'd know exactly what I would get for my birthday and at Christmas. I'm going to have a wonderful time.'

'Oh my,' said the cat, 'now you're getting all your world-lines mixed up.'

'World-lines?'

'Yes, just think how you could make a little map of the route you take to school each morning. But you could also make a map of the route you take through time, or better still through space and time together and that would be your world-line,' said the cat. 'But the important thing is not to get it tangled up or knotted otherwise you'll meet yourself coming round the corner or trip yourself up coming down the stairs. That's what happened to poor Humpty Dumpty. He was sitting on the wall minding his own business when he suddenly saw himself running down the road. It gave him such a shock that he just went to pieces.'

'Oh, I'll be very, very careful.' said Alice.

'Now about that mouse. I don't suppose you *do* have one concealed on your person? I'm feeling very peckish.'

But at that point Alice's tummy gave a very loud rumble. She blinked in embarrassment and suddenly found herself on the street of a small beach town. She walked past surf shops, swimsuit stores and motels, looking for a place to eat. She noticed, sandwiched between a thrift store and a tattoo parlor, a ramshackle building labeled FISH & CHIPS. It was plastered with signs:

DON'T MISS YESTERDAY'S GRAND OPENING! OUT TO LUNCH FOR YOUR CONVENIENCE. COME BACK TO-MORROW FOR A MEAL YOU WON'T FORGET!

The door wasn't locked, so Alice slipped inside. It appeared to be an ordinary café with tables and chairs and a serving bar, but no one was around. In the back she saw three doors. The right and left doors were marked MEN and WOMEN with the usual blue-and-white gender symbols. The middle door, however, was marked OYSTER QUADRILLE: SPECIAL PERSON ONLY, with a blue-and-white octopus sign. Curious, Alice pushed open the middle door and found herself in a small anteroom in front of a desk. The room was decorated with pictures of dead presidents and other serious, important-looking men. A security guard with the octopus insignia on his sleeve was asleep at the desk. Behind the guard was another door labeled OYSTER QUADRILLE: NOTHING REAL CAN BE THREATENED. Alice quietly sneaked past the guard and opened the second door.

This time she found herself in a room full of computers and video apparatus. All of the equipment was marked with the blue-and-white octopus symbol. After a moment, Alice noticed two old men, one dressed in a houndstooth jacket, the other in a typical academic tweed, playing cards at a table. The over-sized deck they were using was printed with colorful pictures of exotic scenes captioned with runic text and mathematical equations. The two men continued playing cards, ignoring their guest.

Not wanting to be rude, but knowing no other way, Alice cleared her throat to get their attention. 'Excuse me, sirs, but can you tell me where I can find something to eat?' she inquired.

Both men looked up, surprised at this unexpected intrusion. Then one of them smiled with recognition and turned delightedly to his companion. 'Oh, how wonderful! They've sent us a lab assistant. Now we can finally get started.' Then, turning to Alice, invitingly, 'Come and sit down. We were just about to order tea.'

Alice joined them at the table as they set aside their cards. The second man inspected her, then said, 'I do hope you are familiar with hyperspace microscopes, for when it comes to machines, we're both all thumbs. Let me introduce myself. I am Dr. No, Professor of Mechanical Metaphysics at Mammoth State National University.'

As Alice absorbed all this, the other man chimed in with: 'He is No and I am Dr. Yes, Professor of Psychological Philosophy at the Splendidly-Endowed Graduate School of Advanced Studies. You must be Tweedle, our new lab assistant.'

Alice realized that they had mistaken her for someone else and attempted to correct them.

'Pleased to meet you brainy guys. But I'm not Tweedle. My name is Alice, and I don't know a thing about machines either.'

'Well, I'm sure you can learn,' said Dr. No. 'We're all beginners here at Oyster Quadrille.'

'But before you start work, you'll have to get a security clearance. This is all Top Secret, you know,' cautioned Dr. Yes. 'We haven't been cleared ourselves yet, so we barely know what's going on.'

'Oyster Quadrille?' queried Alice, confused. 'Top Secret? What's going on here? All I want is something to eat.'

'Oyster Quadrille,' whispered Dr. No in a mock-conspiratorial tone, 'is the code name for a super-secret government project to discover how to manipulate REALITY ITSELF rather than just manipulating mere appearances. Oyster Quadrille is hiring dozens of philosophers because they think we know more about reality than anyone else.' Dr. No continued with more than a little pride for his profession. 'That's what we've been telling people for centuries, and finally Oyster Quadrille is taking us seriously. So this place and others like it are filling up with philosophers—all the big names in the field—but the top Oysters won't let us do any *real* work until we pass our security checks. I'm getting paid ten times what I got at Mammoth State.'

Somewhat skeptical, Alice asked, 'But how can they keep this a secret? Won't people get suspicious when dozens of philosophers leave their jobs at the same time?'

'Oh, officially we're still at our universities. The government hired replacements to teach our classes,' enlightened Yes. 'My Plato symposium is being taught this semester by an expert on Plutonium chemistry. And Dr. No's classes on ethics are run by a nerve-gas scientist and a laser death-ray specialist. They all say they love teaching philosophy and the students benefit from contact with men who've had experience in the *real* world,' Dr. Yes went on.

Then suddenly remembering himself, and impatient with Alice for having distracted them all from their mission, Yes said, 'But we're all hungry! Let's talk over tea. Garçon!'

Alice was so fascinated by the discussion that she hardly noticed when a robot appeared with a tray and served tea and cakes to the three of them. As the robot served tea, Dr. No told Alice more about their work.

Both Dr. Yes and I specialize in quantum reality research—how this world we see comes into existence out of pre-existing waves of mere possibility.'

'Quantum reality? Oh, that's what Professor Who was working on, too,' added Alice knowingly. 'Do you know Who?'

'Everybody knows Who,' Yes responded with a smile. 'Who is mad, crazy, over the edge. Something strange happened to Who in the Black Forest back in the summer of '25. He was a good scientist until then—an original thinker. But now, all is lost. Who actually believes that he experienced REALITY ITSELF in some kind of drug trance. Now he's trying to build some machine inspired by his schizophrenic visions. Who's not a scientist anymore. He's a crazy mystic.'

Doctor No interrupted, 'We're investigating reality too. But we're philosophers, not mystics. We use logic, not drugs, to get to the bottom of things. You've heard of the quantum measurement problem, have you not?'

'Possibility waves when you don't look, actual particles when you do look. So what is looking? Would that be the measurement problem? asked Alice.

'I could not have said it better!' exclaimed No. 'What is looking? How, when, and why do the world's quantum possibilities turn into actualities? I think it's done with mousetraps,' informed No, turning quickly to watch Yes' reaction.

'Mousetraps?' asked Alice, totally confused

'Mousetraps,' responded No, matter-of-factly. 'That's my name for a measuring instrument. Here, look.'

At this Dr. No picked up a remote control and pushed a button. One of the video screens popped on and a cartoon of a machine appeared. A strange, mechanical sounding music blared along with it.

'A mousetrap is a machine in an unstable situation, that can be triggered to jump into some more stable state by the action of a single quantum particle,' shouted No above the mechanical din, as the cartoon illustrated his points. 'When the machine 'snaps' it makes a record that tells us that the quantum is/was there. Your eye is a mousetrap for photons; your ear traps quantum particles called phonons. Physicists build expensive mousetraps to catch quarks, gluons, mesons and other exotic particles, fishing these particles out of the sea of quantum *potentia*—so to speak. Everything we know about the world we must put together from the readings of such machines,' concluded No. 'Mousetraps are our only windows into deep quantum reality.'

'Dr. No was a student of Professor Niels Bohr in Copenhagen and believes, like Bohr, *there is no deep, hidden reality*,' Yes interjected. 'Bohr believed that quantum theory was not a model of the world but only a map of the relationship that exists between measuring machines,' Yes pointed at No, 'which he calls mousetraps—and the world's quantum underpinnings. Quantum theory is not about the world at all but about the responses of instruments we use to observe it. And there's no way—Bohr says—to pull the theory apart to get a clean picture of an uninstrumented world. That was indeed Bohr's philosophy, but I think that's the wrong way to look at things.'

'Dr. Yes was a student of John von Neumann at Princeton,' lectured No, condescendingly. 'He believes that the world is created by lots of little minds, a world in which *consciousness creates reality*. It is a fairy-land of ambiguous maybes of pure possibility in infinitely dimensioned Hilbert Space until some little mind comes along and decides to promote a part of it to the status of full actuality.'

'The world created by minds? That doesn't sound much like physics to me,' decided Alice.

'My teacher, von Neumann, was one of the most brilliant mathematicians of the twentieth century,' defended Yes, picking up a remote of his own, flicking on his own cartoon on another screen, and turning the sound on No's off. Soothing music flowed softly from the loudspeakers. 'Von Neumann showed that if you take quantum theory 100 percent seriously, you cannot avoid a mind-created world. It is a very simple argument.'

'It is a very simple-minded argument,' stressed No. 'Dr. Yes believes that taking quantum theory seriously means that everything in this world—including my mousetraps—must be described by quantum theory,' No said, pointing at Yes' screen, 'and that means that, when you don't look at it, even a mousetrap exists as waves of possibility—the possibility of having "snapped" coexisting side by side with the possibility of "not having snapped." If you buy this way of thinking, then every piece of matter in the universe exists only as waves of possibility,' No concluded, as he clicked on the sound of his cartoon, creating a raucous dissonance.

'Yes, No! I could not have put it better myself!' said Dr. Yes.

Alice concluded that Dr. Yes and Dr. No were working themselves up into a familiar, long-standing debate. Over the clamor of the dueling TVs, Alice could hear Dr. Yes continuing to lecture.

'And no amount of possibility can ever by itself turn into an actuality. By itself the world of matter would have remained forever only a dream world—a world of pure possibility. To make the dream *real*, something must be added from outside, something that is not made of matter, something actual, not possible, that lives outside of quantum rules.'

Dr. No had been losing his patience. Finally he erupted. 'Mind, consciousness, sentient beings, that's what he means! No, no, no, Yes. These subjective realities have no place in physics.'

Turning his back on Dr. No, Dr. Yes addressed Alice, stiffly. 'Despite what No thinks, a mind-created cosmos is the only kind of world logically consistent with the quantum rules. Dr. No's mousetraps, for instance, do they have definite attributes when they are not looked at?'

'Of course they do,' sputtered No. 'Mousetraps are always so big that they behave like classical objects whether looked at or not.'

Dr. Yes' eyes sparked as he turned on his colleague. 'Does an atom have attributes whether looked at or not?'

'No, of course not! An atom is a quantum object, and we all know quantum objects are not things.' No sounded insulted at being asked such a question.

'But aren't your mousetraps ultimately made out of atoms? At what point does a collection of quantum atoms turn into a classical object?' Yes asked triumphantly.

'You don't understand, Yes,' said No, now completely aggravated. 'Objects come first. This teacup, this saucer, these physical apparatuses,' he gestured wildly, pointing at the equipment scattered around the room, 'are certainly real! Atoms and molecules are certainly real also, but have a different kind of existence than this teacup,' insisted No. 'Quantum theory is a mathematical bridge between two radically different kinds of reality—the classical reality of mousetraps and the quantum reality of molecules!'

Dr. Yes had become as worked up as Dr. No. No waved his remote control around and his white hair shot out at every angle.

Why, he looks rather like an electrified Albert Einstein, Alice giggled to herself.

'Why must we divide up the material world into two sorts of realities—quantum and classical?' implored Yes. 'I believe the world of matter is one sort of thing. And the world of mind is another. Mind is what actualizes matter's quantum possibilities.'

Alice had had just about enough of this, and decided to try and settle the argument.

'Is there some experiment you could imagine doing that might settle this dispute?' inquired Alice. 'An experiment that could distinguish whether mind or mousetrap makes the measurement?'

'Every experiment in physics always involves the triggering of some mousetrap. There is no other way to do physics,' Dr. No said with finality.

'But also, every physics experiment at the end is witnessed by some mind. Physics experiments can be automated, but at the end the results must enter somebody's mind, or the results don't exist. There is no other way to do physics,' declared Yes with equal finality.

'So what you both are saying,' Alice said slowly, thinking this through, 'is that because every experiment must involve both a mind and a mouse-trap, then no experiment can ever decide which one is essential to making a quantum measurement? Professor Who is searching for some kind of experiment that can tell the difference, but you think he is crazy to do so, right?'

'Yes, that's right. Who is crazy,' No answered. 'No experiment can ever possibly decide. But it's clear to me that there is only one right way to think about a quantum measurement. Mousetraps make the world real!' With this, Dr. No turned up the volume on his cartoon.

'You're wrong, No, minds make the world real,' turning up the volume on his.

'No, you're wrong, Yes. Mousetraps!'

Higher volume.

'No, No. It's Minds!

Louder volume.

'Mousetraps!'

'Minds!'

'Mousetraps, Yes, you stubborn pig!'

'Minds, No, you ignorant goat!'

'Please, please, this is no way for scientists to behave!' Alice, hands over her ears, finally yelled. She seized the remote controls from the scientists, and turned off the TVs.

'Can't you two just agree to disagree?'

The scientists looked at each other. The room was silent again.

'Tell me more about the Oyster Quadrille project to control reality,' directed Alice, trying to revive a topic that might help her on her quest. At least it would be more interesting than watching Drs. Yes and No argue so.

Suddenly the door swung open, and the security guard strode into the room, followed by a set of identical twins, arm-in-arm with each other.

'Here are your new lab assistants,' said the guard. Then, noticing Alice for the first time, 'Say, what are *you* doing here? Where's your badge? Do you have a "Need To Know"? This is a High Security Area, Miss! I'm afraid you'll have to follow me...'

As the guard escorted Alice toward the door, Dr. No was heard to ask: 'You mean to say you're both named Tweedle?'

'Yes, but it's easy to tell us apart. My sister always speaks with an accent.'

Already forgotten, Alice left the room.

APPENDIX

It was Niels Bohr who first attempted to explain atomic structures using Max Planck's idea that energy is quantized. His approach pictured the atom as a solar system in miniature with planets rotating in quantized orbits. The young Wolfgang Pauli was critical of Bohr's theory and pointed out to his friend Werner Heisenberg, that it grafted a new idea, the quantum, onto an old idea of orbits. Then, in 1925, Heisenberg came up with his own approach, Matrix Mechanics and a year later Schrödinger proposed Wave Mechanics. Pauli, Heisenberg and Schrödinger each visited Bohr in Copenhagen where Bohr attempted to devise a uniform interpretation of quantum theory which became known as the Copenhagen Interpretation. Nevertheless there were subtle differences between the approaches of each of the protagonists and, as we saw in this chapter, Heisenberg favored the notion of *potentia* which harked back to Aristotle. Thus in the deepest sense a level of mystery still lay at the heart of quantum theory.

5 THE QUANTUM HUSSY

Danah Zohar

In which Alice, inside the Quantum Vacuum, meets the dizzying Electron-Girl who takes up with her many Suitors, all at once, in 'Virtual' Relationships before choosing any 'One.'

Alice's head was still spinning from Yes and No's arguments as she walked out into the open air. 'Mousetraps,' she mused, 'making possibilities actual? What if there were a device that worked just the opposite of a mousetrap? Something that turned actualities back into possibilities? Maybe it could be called an *un-measuring machine*,' Alice closed her eyes and imagined a little gray mouse caught in a snapped trap. Then she pictured the trap slowly opening, the mouse escaping and running across the field. 'Strange,' thought Alice, 'the mouse seems to be running away in all directions at once.' Then she remembered—mice. Yes wasn't there a cat somewhere, a hungry cat who asked if I had a mouse in my pocket? So did I really go into a restaurant and did I really find a laboratory with two silly philosophers? Or did the Cheshire Cat dream the whole thing and am I part of the cat's dream and if so...

Alice blinked then closed her eyes. When she opened them she found herself surrounded by a vibrant scene of multiple things which made her dizzy. Each set of multiple things—trees, houses, blades of grass, *everything!*—was slightly different, but each seemed to be recognizable as variations on a single theme. So, while Alice noticed that there were many versions of a given house, each might have a different shape of roof, a different number of windows, or be a different color.

'This place looks like a fun house Hall of Mirrors, except each mirror is slightly different, and each image is not quite the same as its companions,' thought Alice. 'How strange! How confusing!' she muttered out loud.

Alice was enjoying her kaleidoscopic visions when a teenage girl appeared—in multiple copies, each slightly different from the others.

'Hello,' said Alice, wide-eyed and somewhat overwhelmed by the girly crowd.

She was greeted by a cacophony of voices, all of which seemed to be saying different things, simultaneously. 'Hello.' 'Good-bye.' 'It's nice to meet you.' 'Get lost!' 'Where do you come from?' 'Who are you?'

'Wait a minute!' shouted Alice, clasping her ears with both hands. 'If you all talk at once, I won't understand a word.' Alice then stared very hard at one of the girls, who gradually became more distinct, while all the others faded into dim, fuzzy shadows.

'Be careful, you silly girl!' warned the Distinct Image. 'I don't want to lose all my selves. If you look too closely at me, I'll collapse into just one bit of me.' In the background, Alice could see all the other girls gesturing different gestures, and hear the murmur of their different voices, but she kept most of her attention on the girl straight in front of her.

'I'm very sorry,' Alice apologized. 'Though I can barely understand a word you are saying, at least I can see you clearly now. Perhaps you could tell me who all these people are, and why they so rudely talk all at once. And where am I, anyway?'

'You're quite rude yourself, I think,' returned the Distinct Image. 'You tell me not to say many things at once and then you turn around and ask several questions, all at the same time. When since I'm only one bit of myself, I can only answer one question at a time.'

'I'm sorry,' Alice apologized, 'I'll try to be more organized. I've only just arrived, I'm still quite dizzy, and I don't recognize anything in this place at all. Where am I?'

'You've somehow managed to get yourself inside the Quantum Vacuum,' the Distinct Image replied. 'I don't understand how, because there's only one of you and, as you can see, everything else here is many. People and things are always spread out all over space and time in the Vacuum.'

'You're talking riddles again,' retorted Alice somewhat impatiently. 'I've heard of quanta being spread out over space and time, but not people! And how can this place be called a "vacuum"? It's very full,' she added, finding it difficult to concentrate on her companion when all the other girls kept doing different things, all at once, 'Too full, if you ask me!'

'I admit the Vacuum is very badly named,' conceded the girl. 'As you say, it isn't empty at all. It's really quite full. It's full of *everything*—or at least full of the *possibility* and *potential* of everything. Scientists call it "the ground state of reality," but I prefer to call it the sea of potential. It's the ocean of potential out of which everything that exists emerges. That includes you and me. We are all possibilities in the Vacuum…'

Oh, now I see! I must have fallen into the mysteriously powerful 'empty space' Professor Flow told me about, thought Alice.

'…If you had better manners I could be all my potential selves all at once,' the Hussy complained, 'and then I would be more a part of the sea itself, instead of just one wave upon it. At least one wave is all I am if you insist on staring at me so.'

Alice closed her eyes in exasperation and confusion, and in that moment when her attention wavered, all of the many girls became clearer and the noise from them all saying different things at once became louder. When Alice opened her eyes, the situation returned to the one girl becoming more distinct and more audible as the others faded quietly into a fuzzy background.

'Do you have a name?' Alice asked in a challenging tone.

'I have many names,' the Distinct Image answered. 'As many as I like, really. But most people who visit these parts call me the "Quantum Hussy."'

The girl thrust out her hips in one direction, her chest in another, preened her heavily lipsticked lips and batted her mascaraed eyelashes.

'Well, those people can't be very nice,' Alice replied sympathetically. 'Why ever do they call you that?'

'They say it's my social behavior,' the Hussy observed, 'though personally I don't see anything wrong with my behavior. I quite enjoy myself. Do you like boys?'

'Now and then,' Alice blushed, hiding the truth out of embarrassment at the directness of the question.

'Well, do you or don't you?' pursued the Hussy. 'And what do you do about it? Do you have lots of dates with lots of boys?'

'Sometimes,' said Alice cautiously. 'But it's difficult you know. I really do like several boys at my school, and one in particular, but I always have trouble choosing which one to dance with at parties because I don't want to hurt anybody's feelings. They're all so nice.'

'I don't have problems like that. I like lots of boys and I can go out with them all, all at the same time. I just have several dates at once, or several dances, or whatever.'

As she said this, the Hussy suddenly broke into a dance with a boy on her arm, and all the other girls in the background began to dance with different boys on their arms. The Hussy burst into a chorus of song to Mary Wells' Motown melody, 'My Guy':

I am such a busy girl.
My life is just one mad social whirl, with
My guys, My guys.

I live my lives in Quantumland
Looking for the perfect man, with
My guys, My guys.

Others have to choose,
    and do things one by one
But I am here and there and now and then,
    and having fun, with
My guys, My guys.
One day I will settle down,
    my many selves reduced to one,
But not until I've played the field, with
My guys, My guys.

A quantum date is hard to beat,
    so many men that think I'm neat
I dance around with all in town,
    until I've found the one I know, is
My guy, My guy.

While all the other girls were dancing to the music with every possible guy imaginable, the Hussy stepped off the dance floor to talk to Alice.

'Do you mean,' gasped Alice, 'that all these people are really you and that you have many boyfriends? How is that possible?'

'It's possible,' said the Hussy, 'because I am a possibility. And all these other girls are my other possibilities. It's always like that here in Quantumland. Everything here exists as an infinite number of possibilities.'

What an intriguing notion, thought Alice, new ideas coursing through her mind.

'And you must never stare too hard at quantum things. If you do, all their possibilities collapse into just one puny reality. We can only be all of our selves if you leave us alone.'

'This is eerie,' Alice decided. 'But it's also fun. A bit like magic. It leaves me feeling so common, though. There are so many of you and only one of me. Could I become quantum, too? I can imagine all sorts of people I would like to be.'

'That's just the thing!' congratulated the Hussy. 'Use your imagination. It's a part of you that's *already* quantum. In fact, lots of things about your mind are quantum. Every time you imagine different possibilities, or choose between possibilities, or create new possibilities, you're being quantum. If you weren't partly quantum you wouldn't be human. You would just be like a machine.'

Alice closed her eyes and imagined very hard what kinds of girls she would like to be. She imagined herself being tall, and imagined herself being short. She imagined herself with curly blond hair and with flaming red hair. And as she imagined, the hall of mirrors filled with lots of different images of different Alices, each talking to a different Quantum Hussy.

'I've done it! I've become quantum, too,' Alice exclaimed with a dazed, unfocused glance. 'Look, there are many of me.'

'That's great,' the Hussy reaffirmed, 'though as I told you, you were *always* quantum. Your mind is just like the Vacuum. It is *always* full of possibilities, and it actually works according to the same principles as the Quantum Vacuum, the Quantum Universe. Realizing you are quantum was always one of your possibilities, it's your birthright as a child of the light, and now you can use it to explore the future. You can even use it to make the future happen.'

'I don't understand,' said Alice, though the Hussy's words filled her with strange excitement...

'That's what possibilities are for,' the Hussy explained. 'Possibilities are feelers into the future. Ways of testing out what something will be like. The Vacuum uses all of us billions of possibilities to feel its way in creating reality. Your mind can use all its trillions of possibilities to feel its way in every moment to invent a new you, new dreams, and whole new tomorrows.'

'You mean,' asked Alice, 'that here in the Quantum Universe, people aren't just fixed? Everything isn't just determined for me? I can choose? I can invent myself?' she spread her arms joyfully.

'You can invent yourself,' agreed the Hussy, 'and you can also help invent the world. When you imagine lots of possibilities and when you act on some of them, you help weave the thread of reality. Or at least that's how I like to think of it. When I go out with all my boyfriends, I'm performing a socially beneficial service to the world.' Having reassured herself that her outrageous conduct was thus of some great use, the Hussy resumed preening herself.

'But what will happen if and when I leave Quantumland?' Alice asked, suddenly worried about that possibility. 'Will I have to go back to being just one of me again? Will I lose all this possibility?'

'Never!' the Hussy reassured her. 'Or at least it never has to be that way. Now that you've discovered your quantum nature, you can always carry it with you. Your free mind is with you wherever you go. Your own infinite possibilities are forever part of your nature.'

An orchestra was playing, and the Hussy became very distracted by the call of her own love life and began to fade into her many reflections in the dancing Hall of Mirrors.

'And don't forget, Alice,' the Hussy called out, 'because you're a thoroughly quantum girl, when you finally do decide to settle down and choose your own quantum man, you won't lose yourself. Your separate particle aspect will tend to stay somewhat separate and maintain shades of your original identity while the wave aspects of the two of you will merge and evolve both of you into something entirely new. That'll help the Vacuum to evolve, too!'

Alice's last glimpse of her was of an outrageous young girl kicking up her heels in many dances with many boys, as the familiar cacophony of Hussies' voices resumed in song:

I am such a busy girl,
My life is just one mad social whirl...

6 ONE MIND

Amit Goswami

In which Alice learns how her Deeper Self creates
One Phenomenal World out of billions of Quantum
Possibilities.

Just as Alice's last glimpse of the Hussy faded, the whole scene of dancing possibilities became blurred to the point where she couldn't discern
anything. Finally it became pitch black. Ah! that cat must be around somewhere, thought Alice.

As she peered into the darkness she began to make out the image of a
little man who resembled an Indian Guru walking around in a cavern carrying a lantern. He was wearing a saffron robe and there was a dot painted
between his eyes.

'Hello, my name is Alice, and I'm lost,' she called out to the guru.

'Well, hello, Alice,' the little man said as he warmly held her hand. 'My
name is Goswami. I myself got lost once in the deep recesses of Quantumland many years ago, until I met a guide, a mathematician named John von
Neumann. Von Neumann held out the light of consciousness for me, so I
could make my way through the darkness. You see, Alice, the strangeness of
quantum reality is but one act in the grand play of consciousness. Once you
understand the miraculous nature of the quantum, you will understand the
play of consciousness and how it manifests the material world.'

'I don't understand. Whose consciousness? Do you mean my consciousness?' asked Alice. 'I don't *feel* I create anything, let alone the entire material
universe, even though I did *imagine* myself as lots of different possibilities
in the Quantum Vacuum. You, the Hussy, Professor Flow, you're all claiming I can do impossible things!'

Goswami chuckled and teased her, saying, 'And this from a girl who
was told by the White Queen to practice imagining impossible things! Alice, now close your eyes,' he continued more seriously, 'and imagine that
there are two worlds, two universes. One is the conditioned, clockwork,

mechanistic universe of Newtonville where most people live their lives. The other one is the creative universe, the Quantum Universe, a world of *potentia* that non-locally transcends our conditioned universe. It is beyond the conditioned universe, yet it influences it—and creates the contexts for it. To reach it, you have to take a quantum leap. And then you will directly experience how consciousness creates the material world.'

Alice grimaced. 'I don't think I really like all the idea of taking quantum leaps. I've not had much practice and I certainly can't do it—just by wishing it!'

'Why can't you? Isn't that how you came to be here?' Goswami said, indicating the strange domain in which they stood.

Alice thought for a moment and remembered how she came to be in Quantumland. 'But that was an accident, I certainly didn't wish it. Did I?'

Now Goswami challenged her more forcefully: 'Awake, arise, realize that you are both these universes: conditioned *and* creative; mechanistic *and* possibilistic. You are consciousness—original, self-contained, and yet constitutive of all things. All these worlds—the transcendent and the immanent—are inside you! And, of course, they are also outside you, but that is just a camouflage. Strive to penetrate the camouflage!'

Alice wasn't convinced by Goswami's commanding assertions. She shrugged her shoulders.

'I guess you are telling me to meditate, and get down on my knees, or something. That's easy for you—you're from India! But I haven't meditated in my entire life. I wouldn't know what to do,' she finished rather apologetically.

'I will give you a hint.' And with that Goswami began to pace excitedly back and forth in the cave, gaining energy as he moved. He stopped in front of Alice, having come to some private decision. 'I'll give you a hint,' he repeated. 'Do you know who Pavlov's dog was?'

'Well I know about cats in boxes. But now there are dogs as well.' Alice looked puzzled for a moment then remembered. 'Oh yes, I know. You mean the dog that salivated at the ring of a bell—even without the sight of food—because it had been trained to do so?' asked Alice.

'Yes. Yes. Now stop playing the game of Pavlov's dog—and start playing the role of Schrödinger's cat!'

In that instant a cage appeared in the corner of the shadowy cave. Inside the cage, there was a luminous speck, a radioactive atom, blinking on and off. Also inside the cage was a Geiger counter, whose pointer was registering two readings simultaneously, one superimposed over the other: ATOM HAS DECAYED, said one reading, ATOM HAS NOT DECAYED, said the other. A bottle of poison was in a corner of the cage, over which a hammer was poised, connected to the Geiger counter.

'Yes, I know all about that. The cat told me how it works and he certainly did not like that doctor Schrödinger. I know what happens next, if the atom decays the Geiger counter will point to ATOM HAS DECAYED, the hammer will break the bottle of poison, and the cat will die. If the atom has not decayed, none of those things will happen, and the cat will live happily ever after. So, the cat could be half-dead and half-alive after the hour.'

'But where is the cat, Alice?'

Alice thought for a moment, then brightened. 'In possibility? In *potentia*?'

'Now you've got it!' Goswami confirmed.

'I remember. Professor Flow introduced me to the Wiff! It has the ability to manifest any actuality when it pops from a possibility cloud into a particle when observed. Is the cat a Wiff! which dwells in the possibility world of underlying potentia?'

'Exactly. Two possibilities—dead cat or live cat—exist together in one potential Wiff! And for whom are these events possible?' he probed.

'Any consciousness?' Alice startled herself with her own answer.

'Yes indeed,' Goswami confirmed.

'Do you mean if there are possibilities there must be someone's consciousness that these possibilities exist within?'

'Yes, ma'am. Quantum theory describes the unobserved world as pure possibility—live cat/dead cat, in this case. And possibilities can exist only in consciousness—for consciousness, an observer—to choose from.'

'So consciousness is important in the Quantum Universe. The Quantum Universe exists in my consciousness. Now what?'

'Now ask, who creates actuality—an actual material event—dead-cat or alive-cat—out of the possibility wave?' he asked slyly.

'Does consciousness?' Alice guessed. 'Do I?'

'Don't be so hasty,' Goswami cautioned. 'Are you saying that you—Alice—take a little peek at the cat through a hole in a cage, and the Wiff!-cat becomes either dead or alive? How can you have such power?!'

'You're confusing me again!' wailed Alice. 'If the cat sits in my consciousness as possibility, it's *my* consciousness, and I'm entitled to choose its state, am I not? I'm not doing anything to the poor cat. As an observer, I'm just recognizing one of its possible states!'

Goswami smiled, 'Now you've got it! You're admitting it yourself— you do choose what manifests in material reality!'

'You're putting words in my mouth!' said Alice, irritated. 'But that's what *you* seem to be saying. I just took your logic to its natural conclusion. But I don't believe it, logical though it may be. I've never experienced choosing to make reality happen.'

'Now, Alice—make up your mind. Do you, or don't you choose to make an actuality out of the Wiff!?' he asked insistently.

'Suppose I *do* choose. So what?' she answered defiantly, feeling a headache coming on.

'Then suppose that you and I look through the holes in the cage simultaneously? Oh, but there's one thing I forgot to tell you—I don't like cats. But, I know you do. If I choose the cat to die, and you chose the cat to live, whose choice would count?'

'Stop! You're confusing me again!'

Alice thought for a moment. Then...'Wait a minute the cat could not possibly manifest both dead and alive,' she deduced triumphantly, 'nobody has ever seen such a cat!'

'Exactly! Not even a quantum physicist. But, just maybe my choice counts because I'm the physicist! Maybe you're just a figment of my imagination, eh Alice?' he added devilishly.

'I think that you are a figment of my imagination,' Alice said, but she wasn't really convinced herself. 'I'm the one dreaming Quantumland, not you! This is my dream, Goswami, not yours...'

Goswami shook his head and grinned at her assertiveness.

'We both did, kiddo. We both dreamed Quantumland. But let's return to our paradox—the paradox of solipsism: only one observer counts, the rest of us are figments of that one's imagination. We can argue till eternity to resolve that puzzle until you see that the solution is that there is only *one* consciousness. The consciousness that chooses is unitive consciousness beyond our separate diversity. Our separateness is imaginary, a persuasive illusion!'

'Would you please explain that again?' asked Alice, skeptically.

Goswami obliged. 'Underneath our seeming separateness, you and I—and everybody else—are one.'

Alice remembered what the Quantum Hussy had told her about the Quantum Vacuum as the ocean of potential out of which everything that exists emerges; how as individuals we are possibility waves, waving on the sea of the Vacuum.

Goswami continued: 'To the extent that you identify with your separate self, you don't realize that you choose. And you don't see how you create the material universe.'

'Now wait a minute, Dr. Goswami. Why is it that when I identify with my separate self, I don't realize I am choosing? As a separate self, I just did that with the cat, didn't I? And I was realizing I was choosing, wasn't I, after all?'

Goswami answered: 'In some cases our individual choices are in tune with—let's call it the One Consciousness—and in other cases not so. A fully enlightened being's choices would be precisely identical with the One Mind's choices. The illusion of a self, the ego, the mechanism that thinks of itself as the doer is the main obstacle to being a quantum creator. In the Quantum Self there is no individual ego acting, there is action, but no actor. In the brain, experience has formed a distinct pronoun, an "I." And most of us ignorantly identify with that pronoun,' Goswami concluded. 'Now watch!'

On one side of the room appeared what looked like a behavioral research laboratory. A rat was being trained by a research assistant to follow the right branch of a Y-shaped maze by offering it cheese at the end of the 'correct' pathway. Both Alice and Goswami heard the rat's thoughts, although the research assistant conducting the experiment couldn't hear them.

'Heh, heh, heh. This rat gets it. Always take the right branch of a Y,' squeaked the rat to itself.

On the other side of the laboratory, a little girl was being taught by her parents to use her fork in the 'right' way. When the child succeeded, the parents clapped with glee, reinforcing the child's behavior. The parents didn't hear the child's thoughts, but Alice and Goswami did.

'Whoa, claps, approval! I like that,' the little girl thought. She dropped her fork, and then picked up the identical fork in the 'right' way, again. 'Come on, clap again.' the little girl thought. The parents clapped on cue.

A whole montage of similar scenes flashed on the cavern wall: a teacher trying to promote rote learning, a friend applying peer pressure to conform, and many more.

'So that's how we manifest and become bound by the conditioned universe!' lamented Alice. 'So if I stop identifying with the ego, if I truly be the Quantum Self that I am, then I can see for myself that I am creating the conditioned, material universe moment to moment through my experience. All that you say makes perfect sense to me. But I still don't see a way to go beyond that ego-identity.'

'But you don't have to succumb to conditioned responses,' stressed Goswami. 'When you respond in a new way, when you see a new context for responding from infinite possibilities, you are in the Quantum Self—before any conditioning takes hold—when you are being creative. It's just like this.'

Goswami snapped his fingers. On one side of a room, a teacher was teaching a physics class. 'The message of Newtonian physics is clear: Energy and mass are two different things,' the teacher droned.

On the other side of the room, Alice recognized a young Albert Einstein with a light bulb lighting up above his head: Idea! He muttered to himself: 'What do you know, $E=mc^2$: Mass is Equivalent to Energy, and the one convertible into the other!!'

Goswami continued solemnly: 'Alice, creative discoveries—such as Einstein's—are not the product of rational thinking alone. They require the encounter of your ego and the Quantum Self. Similarly, when you creatively intuit a new context for living, unbounded by the ego, you are encountering the Quantum Self again. This is inner creativity, this is the spiritual journey, the context for all the religions of the world. Have you seen Michelangelo's painting on the ceiling of the Sistine Chapel—God reaching out to Adam? That's the creative encounter!'

'But, what is God? That word is so loaded, I'm afraid. Some people say God is a mighty emperor in the distant sky, separate from us, the King of Kings. But I've never been quite satisfied with that picture,' Alice admitted.

Goswami replied joyfully: 'Some say God is a mystery. Some say, God is ineffable, that which cannot be expressed. I think of God as Being and Its Possibilities with no separation between them. Some call God "All That There Is."'

'But how can I experience a God if in God there is no separation at all, not even a subject-object separation? Aren't we back to the Quantum Vacuum again, the sea of potential of all that is possible?' queried Alice.

'If you want a relationship with God, it helps to imagine that the Universal Quantum Self is God, the God within, from which springs your enthusiasm and inspiration. Then God becomes your friend, your lover, and creativity truly becomes an act of love and joy. When you creatively intuit a new context for living, unbounded by the ego, you are encountering the Quantum Self.'

'But many organized religions seem to be so narrow—they don't talk about inner creativity. They just tell you about rules and sins. They tell you to worship God,' Alice argued, 'but they don't usually give you an experience of Him, or your true Quantum Self.'

'People who are intimidated by spirituality, imagining God to be separate from them—a majestic King sitting in the sky—suits their image just fine. For people of creativity, the path to God is the path of self-discovery and understanding.'

'I still don't understand inner creativity. What does one do?' questioned Alice.

Goswami was silent for a few moments. Then he replied, 'Alice, remember when you said you are the one who dreamed Quantumland…?'

'And you said we both dreamed Quantumland. I remember,' she answered.

'Don't you wonder how that is possible? Both of us dreaming the same thing?'

'Yes, I do. Tell me,' she said eagerly.

'The answer is synchronicity—acausal but meaningful coincidence. Consciousness simultaneously collapses the same actuality in both our brains, in both our dreams. The psychologist Carl Jung and the physicist Wolfgang Pauli wrote a book together in which Jung formulated the idea of synchronicity. I call it "quantum non-locality." Quantum non-locality means that behind our apparent separateness Alice and Goswami are One, interpenetrated, made of the same Mindstuff-Quantumstuff!'

No one spoke for a moment; both stood together quietly alone with their own thoughts. Then Alice broke the silence. She spoke timidly at first, then gained conviction, as she went along, declaring, 'I want to wake up to the creative universe! I want an encounter with my Quantum Self! Whatever name you choose to use, I am not afraid of God. I feel God is my friend, my Source, my substance, sustenance.'

Goswami nodded agreeably. 'This is a good way to inquire into the miracle that you are.'

Then he called out a magic word, and a spiral of light appeared, then behind it, a figure of two hands held out in benediction. At the edge of the spiral was a shadowy figure. It resembled a silhouette of a dancing Alice.

As the Alice-like figure danced toward the light-spiral, the light moved toward Alice, and ultimately absorbed her into it.

Alice called joyfully to Goswami, 'Is that the ultimate purpose of quantum leaps, to merge completely with the Quantum Self—to become one with it?'

'Ultimate purpose? Who is talking about ultimate purpose?' smiled Goswami. 'If all of us identify with the One, the world of diversity ends. But at the same time, we needn't give up our uniqueness, our diversity—we only give up the illusion of separateness! Then we are liberated to love and serve the whole world as our family. Then we are free to take responsibility for our environment—which is us.'

Alice imagined herself as the body of her teacher, as the room surrounding them, as the building containing the room. She expanded and became the earth, sea and atmosphere. Then she took a huge breath and tried to encompass the universe. For a moment she seemed to be Everything That Exists. Then she exhaled and reality dissolved around her.

APPENDIX

The so-called 'collapse of the wave function' is one of the most puzzling features of the quantum theory. John von Neumann was the first to propose a rigorous mathematical approach to the quantum theory. He also pointed out that while a quantum system is described by a wave function, this wave function involves a superposition of many different states. In our everyday world a cat is either alive or it is dead. But in the world of quantum cats a cat could be 50 percent alive and 50 percent dead, or even 99 percent alive and one percent dead. It is only when someone makes a measurement, observes the quantum cat that all those different possibilities collapse and only one is left behind. This is termed the 'collapse of the wave function.' But how does collapse occur? It was this puzzle that, after a long correspondence with Einstein, prompted Schrödinger to present his famous 'Schrödinger's cat paradox.'

A number of proposals were made to resolve the problem. Eugene Wigner, for example, suggested that it was the consciousness of the human observer that collapses the superpositions into a single possibility. As the Cheshire Cat already explained, Hugh Everett and John Wheeler proposed that the universe splits when a measurement is made so that only a single state occurs in each of a number of different universes. As we shall see in the following chapter, David Bohm's approach is that there is no superposition but only a single state at any one time.

7 ALICE IN BOHM-LAND

F. David Peat

In which a Taxi Driver demonstrates David Bohm's Wave-Guided, Everywhere-Real World.

Alice opened her eyes and stretched. 'Oh I had such a strange dream, and you were in it too,' she said to the Cheshire Cat.

'Yes, I know.'

'But how could you know, it was my dream?' asked Alice.

'No, it was my dream; I dreamed your dream for you and even put myself in it.'

'But then was none of it real? What about Dr. Goswami and all those things he told me? And...and none of this is quite making sense to me again. I remember you told me that your owner Mr. Schrödinger put you in a box and then you were sort of suspended between being alive and dead. And then a conscious being observed you and all your possibilities collapsed into one.'

'I don't think I quite said that, I told you I kept cat-leaping between different parallel universes.'

'But that's not what Dr. Goswami said.' Alice countered. 'He kept talking about consciousness.'

'Well after all he had most peculiar opinions about cats, probably keeps mice as pets...or even a dog!'

'But hear me out first. Suppose Dr. Goswami had built a very big cage and he put me in it instead of you. Now if I can collapse your possibilities by looking inside your cage then what if it's me in the cage after all? Can't I simply collapse myself? Isn't that right?'

'Oh, I think I'd be in two minds about that one, Alice' said the cat, looking at her with a strange smile.

'What can you mean,' asked Alice?

'Oh, it's simple really. Haven't you ever been in two minds about something? Well you could be in two minds about yourself and "poof" nothing collapses, it's all nice and confused and superimposed as before.'

'But that can't make sense, can it?'

'Paul Dirac didn't think so; he was always in two minds. You see he was never quite sure if he should address his wife as Mrs. Dirac or as Wigner's sister. It was very confusing for the whole family. Sometimes he didn't speak for a whole day which really confused poor Wigner himself as to whom exactly he was related to. You see sometimes Wigner would take tea with Schrödinger and then he'd be thinking about his sister...or should he be calling her Dirac's wife. The he'd take a peek into the box and he'd see all sorts of different cats. It quite made his head ache, being in several minds at once.'

'Oh, that would be a puzzle, what did poor Mr. Wigner do?'

'He had a good friend who used to look in from time to time and stare very hard at the back of Wigner's head until he was only in one mind... but...'

'But what, dear cat?'

'Well sometimes the friend got a little confused, so he also had a friend... and the friend had a friend. Sometimes the queue outside Wigner's house stretched right around the block. So I think it's best not to bother with that guru you met and just stick to plain old parallel universes.'

'Oh, I'm terribly confused, my poor head aches so,' Alice complained.

'In that case I think it's high time for you to have another dream.'

'But I'm not at all sleepy,' protested Alice.

'Oh no, this is dreaming while you're awake. It's called 'dreaming in the dark.' It's cinema, moving pictures.'

'That's silly,' observed Alice, 'pictures can't move. They're hung on walls. They just stay there.'

'Oh, it hasn't happened in your time yet, but don't worry, it will. Just close your eyes for a moment and see what happens.'

Alice closed her eyes then opened them again to find herself seated in a dark room with lots of other people. Suddenly the wall in front of her lit up and she saw a city with lots of very tall buildings. And then there she was, Alice, on the wall, walking down the street.

'Curiouser and curiouser,' said Alice and the Alice on the screen looked down and winked at her.

ALICE outside the Empire State Building in rush hour. A YELLOW CAB draws up.

YELLOW CAB: Get in, lady. Where ya going?
ALICE: I'm not sure...I only...

YELLOW CAB: Yeah, I know. It's Queens you want. Get in.

YELLOW CAB pulls away and at high speed twists and turns through the traffic, rushing down side streets, up avenues, etc. Eventually ALICE finds herself crossing the Hudson River.

ALICE: Oh dear, I wouldn't like to do that again.

ALICE suddenly finds herself outside the Empire State Building again with the same YELLOW CAB. They head off, this time taking a different route and going over a different bridge.

ALICE: What's going on?
YELLOW CAB: You're in Quantumland lady, what else do you expect?
ALICE: No, it's not.
YELLOW CAB: You telling me, lady?
ALICE: Well, if this is Quantumland, and it's not like any Quantumland I've ever seen before, then who are you?
YELLOW CAB: An electron of course.
ALICE: No you're not?
YELLOW CAB: You sure think you know a lot for stranger, don't you lady? Always contradicting.
ALICE: But an electron's a wave...No it's not, it's a particle. Well, sometimes it's a wave *and* sometimes it's a particle...but it's certainly not a YELLOW CAB.
YELLOW CAB: Ah, you've been talking to those other Copenhagen guys. Take no notice of them; it's all magic wands and soap bubbles in Copenhagen. See, here *we're* more down to earth.
ALICE: Then where am I?
YELLOW CAB: Quantumland, but it's David Bohm's Quantumland.
ALICE: Oh, I've heard of that.
YELLOW CAB: Look lady, you want a hamburger you can bite into, or something that keeps vanishing into a wave every time you try to get your teeth into it?
ALICE: I see what you mean...but then...

All the time YELLOW CAB has been starting out from the Empire State Building, dodging through the traffic and taking different bridges to Queens. Indeed, YELLOW CAB doesn't even look where he is going.

YELLOW CAB: I knew you'd cotton on sooner or later. This is the double slit experiment—except we're taking bridges instead of slits.
ALICE: Oh you do make by head whirl so. The double slit experiment is about...well, it's about particles and waves...and...and...about probabili-

ties...and electrons not having paths...and...

YELLOW CAB: Like I said, lady, all magic wands and soap bubbles. Whoever would have believed it...collapsing wave functions, multiple universes. Hey, look over there...*(points to a cat sitting at a sidewalk cafe sipping white wine. Shouts over to the cat.)* How's it going man?

SCHRÖDINGER'S CAT: Cool man. Like, it's a great gig in Bohm's Quantumland. I mean, man, I was wigged in that other place. Sometimes I was there and sometimes I wasn't.

ALICE: I had a cat once, a Cheshire Cat.

SCHRÖDINGER'S CAT: Yeah, man, he was my cousin. But that stuff about the grin was nothing...ever had your wave function collapsed? And having to spend all that time in the box waiting for Schrödinger to open it up. I mean, man, who needs that. Here, in Bohm's Quantumland you can sure bite into the fishburger. Ciao, man. See you around!

YELLOW CAB drives off.

ALICE: I'm beginning to see that things are a bit different in Bohm's Quantumland. But then if you're a real electron and not a wave then how do you get around. All I knew about physics at school were pushes and pulls.

YELLOW CAB: That's old mechanical stuff, lady. None of your old-fashioned mechanical stuff for Bohm. See this. *(He pats a radio control box on the dashboard.)* I use the good old Quantum Potential.

ALICE: You mean...?

YELLOW CAB: It guides me around the city. Watch... *(YELLOW CAB flies up into the air so that ALICE can look down on the city. She sees other YELLOW CABS taking detailed paths along streets and avenues.)* Look, lady, the Quantum Potential knows about the whole city and we electrons sort of plug into it.

ALICE: You mean the Quantum Potential guides you all.

YELLOW CAB: You got it, lady.

ALICE: Then it must be very smart.

YELLOW CAB: Sure. But what about me? I have to pick up the information. You thought electrons were just little particles and waves—well we're not. We're smart, old Bohm once said we were inexhaustible in our richness. 'Inexhaustible,' I sure like that. Like I say, we're real smart; we *surf* along on the quantum potential. We're reading it day and night. That's how we always know where to go. When you're going through those double slits you don't want any of that nonsense about probabilities and indeterminism. Here in the Bohm *world* we always know where we are...and we know where we're going. And I'll tell you something else, lady, see those two cabs. *(From on high YELLOW CAB points down to two cabs on opposite sides of Manhattan.)* Well they're both surfing on the same quantum potential. And you know what? They're both connected. Want to *see* something else? *(He rises even higher so she begins to see the highway pattern of New York State.)* It doesn't even stop here. The quantum potential goes on and on so wherever you are, no matter how far away, we can still read it. It's real smart. And no matter how far those two guys travel they are always connected. We're all connected. Bohm's Quantumland is a holistic world.

ALICE: You're making my head spin again. Let me get this right. The quantum potential knows everything about every street in every town? And you're all plugged into this. You're all being guided by it? *(YELLOW CAB nods.)* So you're all connected together in a way. Everyone's part of a *great* family.

YELLOW CAB: Well, sort of. But less of that 'part of' because there really aren't any parts, just wholes within wholes within one great, exciting whole.

YELLOW CAB is back driving in Manhattan again.

ALICE: But you're a part. I mean I can see you. I can touch you. You may be connected in a way…but…but…Well, you're here and that cab's over there. And, well, it's over *there and you're* not. So you can't really be the same. You are separate. You are parts.

YELLOW CAB: Lady, are you in for a surprise! We don't usually let visitors see too much at first. It's bad for business. But you, you're always asking too many questions. Hold onto your hat.

YELLOW CAB begins to flicker and break apart. He begins to spread out as a yellow wave which expands across the street, across Manhattan, then begins to collapse inward, until the wave becomes YELLOW CAB again.

YELLOW CAB: Did you like that, lady? Well watch this.

YELLOW CAB repeats the process but this time much faster. Each time the wave collapses inwards YELLOW CAB is a bit further along the street. *The process repeats faster and faster until YELLOW CAB simply flickers on and off, moving along the street in tiny bursts.*

YELLOW CAB: Parts indeed! The problem was you weren't looking fast enough before. And we're all doing it…all the time. Most people just look too slow to see the reality behind reality.

Now other cabs, buildings etc., begin to expand into waves and collapse again.

ALICE: So you really are a wave after all!

YELLOW CAB: Depends how fast you blink.

ALICE: But then how...?

YELLOW CAB: How does it all hang together? Like I told you, our old friend the Quantum Potential. See, it's really a vast field of information, information about *the whole* universe.

As they talk the scene begins to transform into scintillating waves that constantly form patterns and dissolve. Images suggest themselves, dissolving into each other—winking lights on a computer, electronic information passing through circuits, a human brain, signals passing along a nervous system, an electronic map of city streets.

ALICE: Well, it's not like information in a book. It couldn't be.

YELLOW CAB: No, it's *active information*. It's sort of intelligent and it keeps everything in its place. Remember all those guys in Copenhagen and those other places? They were always telling you nonsense about observers and multiple universes and all that. But really everything hangs together in a beautiful way so there's no *need* for those magic wands *and* soap bubbles.

ALICE: But...well...it's almost as if the Quantum Potential were intelligent, as if it could think.

YELLOW CAB: The mind of the universe? Well, why not? Maybe mind and matter are not that far apart after all. Maybe the universe does think and your mind is part of the thinking.

ALICE: But then everything... (*ALICE has a sudden flashback to her home, her familiar things.*) You mean all this is just...well, a sort of illusion? I don't think I'd like that. Is that what David Bohm is saying?

YELLOW CAB: Oh, it seems real enough. You can sure bite into the hamburger in Bohm's world. But in another way it is an illusion, but a very clever one. David Bohm calls the world you're used to 'Explicate Reality—a projection of what's enfolded inside.' Everything you can touch and see and feel is the Explicate Order.

ALICE: But then things are really...

YELLOW CAB: Underneath they're really like this. (*He dissolves into a wave and returns.*) Or in a way they're both. When you're in the Explicate Order then the things around you are really real. But when you in the... Say, why don't you come with me this time?

They both dissolve. Alice becomes a wave but we can still identify with her...maybe through her voice or just her eyes.

ALICE: Where are we?

YELLOW CAB: We're in David Bohm's Implicate Order. This is the ground that underlies all reality. It's the ground that gives birth to the Explicate Order—look... (*They unfold back to Explicate Order for a moment then enfold back again.*)

ALICE: But where are you?

YELLOW CAB: Inside you, (*a sort of flickering Al ice with the Yellowness of the cab inside*)...and at the same time you're inside me... (*Yellowness with Alice inside*). We're both enfolded, we're one inside the other.

They unfold back to the Explicate Order.

ALICE: This is making my head spin.

YELLOW CAB: Well, look at that. See that building way over there... (*He points to a skyscraper in the far distance. They enfold into the Implicate Order and the flickering skyscraper appears inside Alice and Alice appears inside the building.*) Everything is enfolded in everything else. Everything connects. That's David Bohm's world. The whole universe is one giant, intelligent movement. It's a sort of dance, a ballet, a piece of music. Some people even say it's like a holograph.

ALICE has a vision of the universe in terms of art, music, dance, etc., swirling, insubstantial images, a holograph. Then finally she's inside the YELLOW CAB who stops outside the Empire State Building and opens its door.

YELLOW CAB: Enjoy the ride, lady? See, things are always more exciting than you think. In Bohm's world there's no need for Quantum and Classical, and there's no distinction between matter and mind because it's all part of the movement. Now don't you forget that when you're talking to those other guys—tell them to go with the flow.

YELLOW CAB begins to drive off, talking to himself.

APPENDIX

The physicist Paul Dirac was noted for his eccentricities. Once, after a seminar, he asked if there were any questions. Someone in the audience mentioned that he had difficulties in following the first part of Dirac's argument. After a long pause the chairman of the seminar asked Dirac if he would answer. 'That was not a question,' Dirac replied.

Dirac married Margit Wigner, sister to the famous Hungarian theoretical physicist, Eugene Wigner. It was a running joke amongst physicists that he referred to her, not as his wife, but as 'Wigner's sister.'

8 THE CENTER OF THE UNIVERSE

Brian Swimme, Nick Herbert and William Brandon Shanley

In which a Cook named Beatrice
shows Alice the Cosmos.

Alice opened her eyes, 'That made my head spin. I don't much care for moving pictures. They certainly won't catch on.'

The cat smiled at her and the rest of him vanished.

'Oh please don't do that again. I want a whole cat, not just a smile. And what's more I never even had a bite of that reality hamburger!'

'Still hungry, are you? Maybe you need to meet the cook. Blink once.'

Alice blinked and found herself in front of a storefront with a sign over the door: THE CENTER OF THE UNIVERSE. She blinked again but now the sign had changed. It now read: FOR LOST SOULS ONLY. Alice walked inside.

It was a greasy diner, only one table was in use, and a couple of people were at the counter drinking coffee. There was no waitress, just a cook, a nondescript woman in her fifties wearing a soiled apron, seated at the counter next to the others. Alice took a stool next to her.

'Hello, ma'am. I'd like to introduce myself. My name is Alice.'

The cook turned to greet her. 'Pleased to meet you, dearie. I'm the cook around this place, but "Beatrice" will do just fine.'

'Nice to meet you, Ms. Beatrice. Is this diner really the Center of the Universe?' Alice asked.

'I said "Beatrice" will do just fine,' retorted the cook. 'And just who are you to be asking if this really is the center? Are you really a Lost Soul?'

'I don't see how that matters. I am a little lost...I would love to go home. But I asked a very simple question and I don't see how the condition of my soul should affect your answer one way or another,' stated Alice mulishly.

'Well, never mind,' the cook reflected for a time, then: 'To answer your question then, yeah. This is the center.'

The cook said nothing more, and while Alice's eyebrows rose in anticipation, the cook calmly shook salt from a shaker onto the vinyl counter and made a small pyramid. Then the cook turned to Alice and looked at her, waiting for a reaction.

'Is that all?' Alice erupted when no more information seemed to be forthcoming.

'What more do you want?' asked the cook, matter-of-factly.

'But I don't see anything different!' complained Alice.

'Oh, you want to see *stuff*! Why didn't you ask for that in the first place?'

The cook slammed her palm onto the counter, and in a flash of light she and Alice were suddenly suspended in space, surrounded by stars. Only the two of them remained; Alice was still seated on the stool, the cook leaned on the counter, and all of New York City, even the diner, had disappeared. Alice gasped. She looked at the stars in wonder. They were everywhere, and very bright. The band of the Milky Way cut diagonally across the sky in front of her.

'Oh, my! Where are we now, Beatrice?' Alice asked in awe.

'You've been learning that everything is particle and wave?' the cook asked Alice. When Alice nodded her head in the affirmative, the cook continued. 'I just put our nearby terrestrial surroundings into a wave state. We're in the same spot, but now we can get a better view of our cosmic surroundings. So have a look…' She pointed. 'There's the nearest star, Proxima Centauri. It's four light years away.'

'How far?' asked Alice, very surprised.

'The light you're seeing took four years to get to your eyes,' the cook explained. 'So when you were ten, the light left Proxima, and during all those years you went to school and practiced ballet, the light soared through space 300,000 kilometers, or 186,000 miles, each second. And it still took four years to reach you.'

Alice was surprised Beatrice knew anything about her, but her astonishment at what the cook was saying about light was *overwhelming*.

'I know what you're saying, but that's so hard to comprehend,' Alice admitted.

The cook continued her spectacular tour of cosmic magnificence. 'See these four hundred billion stars from the Milky Way Galaxy? You said you want to go home? *This* is home. This is where you were born, this is where you developed, this is where you will return,' the cook added, somewhat profoundly.

'This Milky Way Galaxy is the Center of the Universe?'

'Yes.'

'But how do you know?' questioned Alice.

'Oh! So now you want proof?' Beatrice was incredulous.

'Well, yes.' Alice didn't think this was an unreasonable request, but the cook was quite flustered.

'You keep changing your questions! Why do you want proof?'

'People have been saying the strangest things to me in Quantumland. If you want me to be honest, I have to say I don't know who to believe in this strange place. So I don't know whether to believe you or not.'

'Do you think I care if you believe me or not?' challenged the cook.

'But if you have knowledge of the universe, shouldn't you want to teach it to people?' asked Alice, curious now.

'Oh, I see. You think acquiring more knowledge is always a good thing,' the cook said with a somewhat contemptuous tone in her voice.

Alice was puzzled. 'Well, isn't it?'

'It doesn't matter what state your soul is in, is that what you think?'

'There you go about my soul again, and all I'm trying to understand is how the Milky Way Galaxy is the Center of the Universe,' she declared in frustration.

The cook took a deep breath, hit the counter again, and the stars of the Milky Way vanished, only to be replaced by the structure of the large-scale universe. The cook took another deep breath, and commenced her discourse.

'If we put the Milky Way stars in a wave state, we can see right through them and then look at the other galaxies. Over here are the two Magellanic Galaxies, the nearest ones to us. But here's the Andromeda Galaxy. It's just about the size of the Milky Way, two million light-years away. Just think what that means. The light that reaches your eyes right now left Andromeda when the first humans were evolving in Africa. For two million years this light rushed through space and all that time humans developed their minds and their civilizations, so that just now we have broken through to an understanding of our place in space and time.'

'Wow! That's the farthest galaxy from us?'

'The outermost edge of the universe is fifteen billion light-years away,' the cook sagely added.

'The outermost edge...And we're at the center?' asked an astounded Alice.

'Yes.'

'And the edge is fifteen billion light-years from us?' she asked, for the sake of clarification.

'Yes. And the proof comes from the fact that the universe is expanding in all directions. The galaxies are not just sitting there in space. They are all moving away from us. This expansion was predicted by the field equations of Albert Einstein, and confirmed by the observations of Edwin Hubble.'

'Moving away from us here?'

'Watch. Here's what the universe will do over the next five billion years.'

The cook directed Alice's attention to the space around her. As Alice watched, the galaxies all quickly expanded away, leaving a much more sparsely populated night sky.

'Notice that all of the galaxies move away from us,' the cook said. 'None of them come any closer to us, because we are, as Hubble discovered, at the very center of this expansion. But now, watch this. Let's move backwards in time, first going back to where we started, and then moving yet another five billion years back in time.'

The galaxies all came back to the way they were at the beginning, and then they moved into a denser night sky, all quickly done, in the same style as the hyper-space speed in the movie *Star Wars*, which Alice had seen at the theater.

'Should we keep going back in time?' asked the cook, breaking into Alice's reverie.

'Oh, please, yes,' responded Alice eagerly.

The stars continued to crowd until Alice, the cook, and the counter all dissolved into the hot gas of a star. Then the atoms of the star broke down into particles, and even those broke down into quarks that flashed on and off.

'Just like the Yellow Cab!' thought Alice.

The cook ignored this last, and continued to narrate during this oddly beautiful chain of events: 'As the galaxies crash back into the Milky Way, all the structures are broken down, and we have only stars left. And then the stars are destroyed, and the universe becomes nothing but atoms. But as the collapse upon us continues, even the atoms are broken down into the tiniest particles, the quarks and leptons. So here we are. At the beginning of the universe. With all the trillion galaxies crashed in upon us in one tiny volume of space, smaller than a mustard seed.'

Alice was feeling quite queasy; frightened by the visions Beatrice had used to demonstrate her points.

'Thank you very much, I do appreciate this journey, but could we go back now?'

'You're not enjoying this? I thought you wanted proof?' said the cook, testily. 'Don't you have any questions?'

'Well, yes, I most certainly do. How is it possible that a universe thirty billion light years across could shrink down to a size smaller than a mustard seed?' Alice asked with more than a dash of skepticism.

'No one knows the answer to that question, dearie. At such tiny distances, physics as we know it ceases to apply. Physicists call this mustard seed the "Ultimate Singularity" but they could equally well have called it the "Ultimate Mystery." It's where everything began and it's the proof you were looking for.'

'This is not what I asked for, and don't pretend that it is. I wanted proof that your little coffee shop was the Center of the Universe, and here you have brought me back to the beginning of time!'

'But we're still in the coffee shop, that's the whole point,' said the cook with great aplomb.

'We most certainly are not!' argued Alice.

'In one sense we're not, of course,' elaborated the cook. 'But we haven't moved from the coffee shop. This is the same place in the universe that will later become our coffee shop. Don't you see?'

'So we've just gone back in time?'

'Yes! Watch.'

And with that, the cook sent everything into reverse. The particles assembled into atoms again, which in turn started to swirl as a star's gas. Then the stars reappeared in a great density that thinned out until the two of them were sitting at the counter again, surrounded by the night sky all around.

'First the particles came forth fifteen billion years ago, then they formed atoms, then stars and galaxies, and now us,' the cook explained. Then, with another slap on the counter, the diner and New York City re-materialized around them.

'And so you're telling me that this diner of yours, this—excuse me—*crummy* little diner is the center of the whole blooming universe?' scoffed Alice.

'Here we go again,' sighed the cook, rolling her eyes.

'I'm sorry, but of all the ridiculous ideas…'

'This one takes the cake,' said the cook.

'…this really takes the cake,' said Alice.

'It's so arrogant…' began the cook, in an undertone, 'to think the human is at the center.'

'…to actually think the human being is at the center…' Alice was saying at the same time.

'But no one here is saying that!' the cook interrupted Alice's skeptical tirade. 'I'm simply telling you that this place is the center. And it's even harder to understand than I've indicated. You know, I sometimes really wonder why I bother.' This last was directed at no one in particular.

'Now don't get upset with me,' demanded Alice, rather unreasonably after all she had put the cook through.

'You come to me, you ask me about the Center of the Universe, I tell you what humankind has learned in twenty-first century mathematical cosmology…'

'And I do appreciate it, really,' interjected Alice, now sheepishly.

'…I tell you all I know, and then I have to deal with all your hang-ups. You think this is enjoyable?'

'I am sorry. Please. Do you think this is easy for me?'

They were suddenly switched back into the night sky. It happened so abruptly that Alice was left breathless.

'OK, look, there's the Coma Cluster. It's a billion light-years away. You want to see what the universe looks like from there?'

Before Alice could respond, there was the familiar yellow flash Yellow Cab had used to move about in Bohm-land. They flashed to the Coma Cluster, and reassembled as Alice and Beatrice, but there were some major anatomical differences, representing the evolutionary differences that would take place in the future. Once they arrived there, they went through the same journey into the future and back in time as they had in the Milky Way. It took only a couple of seconds each way, this time. After the Coma Cluster, they jumped to the Perseus Cluster, and Beatrice took them through the same sequence again, as she explained.

'Does it look familiar?' she asked. 'Now watch what happens as we move forward in time. Do you see how all the galaxies are moving away from us?'

'Why, yes. Yes, I do,' Alice marveled.

'And why? Because Coma, just like the Milky Way, is at the Center of the Universe. And, as we go back in time, we also see that Coma is where the universe began,' the cook explained. 'Now, let's try it again from the Perseus Cluster.'

Their location changed at hyper-speed to the Perseus Cluster. The cook continued her cosmic tour: 'See, Alice? So it really doesn't matter what point in the universe we choose. As we move forward in time, we will see all the galaxies moving away from us. And as we move backwards in time, we will see all the galaxies collapse upon us, at the unmoving Center of the Universe.'

'So there's lots of centers,' Alice deduced.

'Yes! Even more than lots,' said the cook excitedly. 'It means that every place in the universe is the Center of the Universe.'

'And every place is also where the universe began?' Alice offered.

They flashed back to the coffee shop. The cook folded her arms. 'You've got it,' she nodded.

'So where I'm standing right now, this is the very place where the universe started.'

'That's right,' said the cook with approval.

'All the particles, way back at the beginning of time, this is where they emerged,' recited Alice.

'From this very spot,' grinned the cook. 'The quarks erupted and some eventually became this salt, and some eventually developed into this Formica, and some journeyed through time to become you.'

Alice paused, turned around, and glanced at the door. She thought that she had heard someone calling her name. Nervously, she said, 'Tell me quickly then, where did it all come from?'

'You already know the answer—from the Quantum Vacuum, from the Great Void, from the consciousness of the Quantum Self, guided by the Quantum Potential.'

'But these are just words,' said Alice, exasperated.

'Yes, but they point to a reality at the base of the universe,' stated the cook.

'The Void...' mused Alice.

'I like the designation "All-Nourishing Abyss." The quarks, the galaxies, the diner, Alice, all humanity—all of us—are manifestations of the infinite fecundity at the root of the universe.'

Alice heard her name being called again, and turned about, distracted. 'A magical, mysterious, mathematical place, a place where quantum physics reigns...' she said, pondering the implications.

'But there's another way to think about it: as a realm of unimaginable generosity. Existence pours forth out of the All-Nourishing Abyss; the Existence of the Milky Way, the Sun and its planets; the existence of your family and friends. It all springs forth in each and every moment for an eternity from this generosity at the base of being.'

'When you speak, I can almost feel it. But I know this feeling will disappear. Can you help me remember?' asked Alice wistfully.

'Now you are asking the real questions in your soul. You want to remember where you are in the universe?' Beatrice asked. 'Then you must first feel a deep sense of being lost. If you do not feel lost in the midst of the dualistic, mechanistic society of the twenty-first century world, you have been completely captured, kidnapped by the myth of the materialistic camouflage.'

'But I do feel lost. So what then?' prodded Alice.

'You must learn to deal with the doubt that besieges you.'

'What doubt? Doubt about the Center of the Universe? Doubt about the Quantum Vacuum? Doubt that I'll ever get back home?' Alice anxiously inquired.

'Doubt concerning your *place* in the universe. In each moment, you exist at the Center of the Universe, and you arise out of the All-Nourishing Abyss. To feel this should be as natural as feeling the wind through your hair.'

'But it's not,' said Alice.

'Because you doubt your own experience.'

'But I was taught the world is a dangerous, scary place, to watch out, to never take risks,' Alice confided, 'and to never trust one's own feelings or desires.'

'You will experience the world to be a dangerous place if you believe it to be, Alice. You will feel whatever you think or believe. You doubt your *own* experience because you were taught to distrust the desire that is welling up within you—the same desire that is welling up from the Center-Most

Heart of the Universe. The very same desire that has birthed the magnificence of a rose, the sweet, loving tenderness of a mother's touch, and the miracle of the Sun's rise for eons.'

The voice calling for Alice suddenly came very close, and shouted very loud. She started to move toward the door. She stopped and asked one question:

'But how do I deal with this doubt?'

By holding onto your questions!' the cook called after her. 'They are *your* very own search for meaning. If you can remember only one thing, remember that your questions arise from the same fecundity that gave birth to the quarks and the galaxies. By pursuing your questions, the adventure continues!'

Pondering this thought as she opened the door, Alice turned to thank Beatrice, and then she was gone.

9 CHAOSLAND

F. David Peat

In which Alice learns that Chaos is a Subtle Form of Order.

This time Alice thought to herself, Center of the universe. Parallel universes. I think I'm beginning to get the hang of how to do this. Let's just see...

Alice blinked and found herself in an empty field under a high blue sky and coming towards her was a very curious little man. He was wearing a jacked covered with buttons and badges and he kept bending down to pick up bubble gum wrappers, newspaper clippings, plastic bags, and paper clips all of which he stuck to his coat, pants and top hat. When he came closer he nodded to Alice, smiled and introduced himself.

'Good afternoon, Alice. I've heard a lot about you. Now it's time to take your education in hand. It's not a good thing for a young lady to spend too much time in Quantumland. So, come along now, and off we go.'

But Alice stood her ground, 'Go where? You seem to take a lot for granted. We haven't even been properly introduced.'

He merely threw up his hands and said airily: 'Oh, don't worry about *that*. Everyone in Quantumland knows *me*! They call me "Strange Attractor," but I'll let you into a little secret. I'm the normal one and everyone else is...well, just a little stuck in their ways. Come along now, girl, in here. I'm going to show you Chaosland.'

A door appeared out of nowhere and Strange Attractor led Alice through it. There she found a darkened room with a tiny door at the other end. She realized that they must have shrunk a lot because the closer they got, the bigger the door became. Then Strange Attractor opened the door, revealing a bustling carnival Midway outside! In an instant, he bounded out and turned a triple somersault, catching something up in his hand as he did so. Alice realized it was a shiny button that he had found on the ground.

Strange Attractor was fastening it onto his jacket when he noticed that she was staring quite intently at what he was doing.

He gave her a wink, 'Life's a lot of fun, despite what those quantum characters tell you. After all, never take anything for granted, because tomorrow is always a new day. Didn't you find some of those quantum characters a bit serious and pompous, and the others quite mystical and zany?' he smiled conspiratorially.

'Well, I...' stammered Alice.

'Of course you did, girl. That's because they live with their heads in the clouds—or rather, in the waves. But we're going to look at *real* life for a change.'

Strange Attractor stopped and pointed to a carousel. 'Did you ever wonder why people keep doing the same things in life, going round and round, making the same old mistakes, time and again?'

As he said this, a rather pompous looking passerby slipped on a banana peel. Alice blushed in embarrassment for the man but couldn't stop a giggle.

'See what I mean?' Strange Attractor laughed then pointed to a man on the carousel, going round and round and surrounded by a group of women.

'Look at Mr. Smith. Married five times. Each time he thought he was escaping, and each time he discovered he was marrying the same wife.'

'You mean people just keep going round and round? But where's the freedom in that?'

'No, not everyone,' he gently corrected. 'But hang on, what's wrong with going round and round? After all, everything else does.'

As they watched the carousel transformed into the rotation of the Sun across the sky, the waxing and waning of the Moon, the seasons rotating, spawning salmon, and all the other cycles of Nature circling, circling, circling.

'Most things go round in cycles, don't they?' Strange Attractor observed. 'That's what time is all about. Have you ever thought where time would go if there were not circles?'

'Well no, Mr. Attractor, I can't say that I have.'

'Without circles, Alice, the Earth would not circle the Sun, there'd be no sunrises, sunsets, seasons, or any new tomorrows.' Strange Attractor pointed to another attraction on the Midway. This one was a mass of cogs, wheels, hands on clocks, suns and moons rising and falling.

'The universe is circles inside circles,' he noted, then stopped in his tracks, saying, 'and yet, there's more to it than that.'

'Oh, dear. I'm still not sure I'd like to get stuck like *that*. Let's look at something else,' said Alice.

But Strange Attractor shook his head, 'No, first you've got to try to understand it all. You've got to look at the cycles inside the cycles. Come with me.'

He pulled her into the car of a roller coaster. It chugged its way to the first crest, high above the Midway. From this height, they could see beyond the Midway to a landscape of hills, ridges, rivulets, and valleys. Beside them, on the parallel track, the man with five wives appeared. To the hysterical squeals of all five wives, his car plunged down into a circular valley where it went round and round the wall of the valley.

'See, that's how Mr. Smith got trapped with all his wives. These valleys are called 'attractors.' And so Mr. Smith goes round and round in the same circular pattern, leaving one wife and picking up another, making the same mistakes and all the time he believes that he's doing something new.'

'How terrible!' cried Alice.

Strange Attractor smiled. 'Oh, don't worry about him, my girl. He's been round so many times he's forgotten even where he came from! The great thing in Chaosland is to hang onto your story. Never forget your name and address.'

'Oh, no, I won't,' Alice assured him. 'But what do you mean, and whatever are you talking about? I'm lost, now more than ever. I see the Sun, the Moon and Mr. Smith going round and round in circles. But surely not everything's like that?'

'Oh no. Some of the attractors are a bit more sophisticated than circles. You know, wheels within wheels and so there are loops inside of whorls inside of spirals—but it's all the same trap in the end.'

As he pronounced the word 'trap' the roller coaster went over the first hump! Now Alice was squealing in delighted excitement, as she and Strange Attractor plunged down the tracks into a valley, then sped swiftly up to a level plateau. Alice caught her breath and looked around. In all directions Alice could see the images of other attractors and other cycles. She saw cars riding on tracks which were wound around tubes, which were themselves wound around loops coiled up into spiral paths, which in turn were wrapped around huge do-nut shaped frames. There were loops within loops within loops everywhere. And some of the tracks she could see did not stand still but were themselves moving!

Oh, dear, I don't think I like any of this. But then..., she was thinking.

'Yes, girl?' he encouraged.

'They once told me at school that the only alternative to order was chaos. Now, order may be rather boring, going round and round on the same loop. But I certainly would never like to be plunged into chaos!'

'But that's just where everyone got it all wrong. Chaos isn't a bit like what you were told at school. It's just another kind of order in Nature. It's alive, it's exciting and it's free!'

'But...well, it's all chaotic, isn't it?' Alice asked, not really understanding what Strange Attractor meant.

'No, you see chaos is really a different kind of order. It's an order that's...but wait, I'll show you! Up we go again.'

Now the roller coaster car climbed high up to the top, where a very alarmed Alice realized that the rails suddenly went off in opposite directions.

'What's happening?' Alice screamed.

'A bifurcation point. Don't sneeze or we'll go in the wrong direction.'

Obediently, Alice stuck her finger under her nose, desperately attempting to ward off any tickles that could result in disaster. The car hurtled toward a series of bifurcation points, as Strange Attractor shook his head, shouting: 'Decisions! Decisions! Decisions!' He guided the car through the bifurcation points by leaning, twisting and turning his body, as the bifurcation points came faster and faster and faster.

'Hang on, we're getting close to my world...the world of chaos.'

Abruptly the tracks ended and the car shot up into a nebulous cloud. Strange Attractor looked at an alarm clock he was carrying, although Alice couldn't say just when he had begun to hold it.

'And half past quarter to... Time to get to work.'

'Let me look,' Alice asked, curiously.

He tilted the clock so Alice could see. She stared at its face, whose hands did not go round, but spun, jerked and danced in clockwise and counterclockwise directions, seemingly all at the same time. Strange Attractor put the clock back into his coat pocket saying, 'Don't delay me, girl, or I'll be early.'

Then he split apart into a myriad of tiny Strange Attractors, which in turn split apart into even more Strange Attractors.

A large bumble bee appeared amidst the flurry of attractors raining down all around. 'Good afternoon, my dear, Alice,' the bee greeted her from the center of the flurry.

'Well, good afternoon. And just who on earth are you?'

'Bumble is my name. I'm after the honey,' the bee said as it danced a very erratic dance through all the copies of Strange Attractor.

'Oh, my head's starting to spin again! Just *what* is going on?' Alice demanded.

'The thing is, you've come from the world of order. It must seem very not-at-all-strange to you,' said Bumble, nonsensically.

'Not-at-all-strange? What do you mean? It's certainly not at all like the world of order. It's very strange!' she said with exasperation.

'No, I'm not,' said Strange Attractor's voice. It seemed to come from everywhere at once.

'Shall we show her how it's done?' Bumble asked Strange Attractor.

'Oh, very well. Hold on while I reconstitute,' he sighed. He collapsed back into his original form.

'Now, I'm after the honey, my dear,' repeated Bumble.

'And I'm the honey pot,' said Strange Attractor.

'Remember all those boring cycles and attractors?' Bumble asked Alice, as she circled around Strange Attractor. She started to do loops within loops. Faster and Faster until all Alice could see was a sort of dance of chaos.

'That's called "period doubling to chaos,"' said the bee. 'I split myself in half and then those halves split again and if I keep on doing that I find myself in so many different places at once that it all ends up in chaos and I never reach the honey. So it doesn't really get me very far. That's where Strange Attractor comes in.'

The bee turned to Strange Attractor asking, 'Would you mind?'

'At your service, Madame,' he replied politely.

Strange Attractor began to split up again into his fractal dimension... but this time more slowly, and in stages. Bumble followed him, trying to go from image to image, saying, 'See, that's where chaos begins...trying to reach the honey.'

'I see,' said Alice. 'Then it really isn't...?'

'No, it's not chaotic at all,' Bumble said anticipating her question. 'It's a complicated dance that goes right to the heart of matter. It's the dance of the universe, the dance of freedom. Come with me.'

Alice felt a tingling in her back, and saw that she was growing wings. She also noticed that Bumble seemed to be getting larger and larger, until she realized with a start that she was actually getting smaller and smaller, until she was the same size as the bee. Together they danced from image to image...then together they began to shrink, even smaller until they entered images within images within images.

'You see, dear? Strange Attractor goes on and on, smaller and smaller. The universe is a sort of mirror that repeats and repeats at smaller and smaller distances. That's what "fractals" are. Look...'

Bumble reached forward, grasped a mirror and handed it to Alice. She looked into the mirror and saw her own image—as it broke apart into fractals. Then she saw coastlines, trees, river deltas and blood circulation systems, all at higher and higher magnification and showing self-similarity. She saw a beating heart and its electrical record. She zoomed into the record and saw its fractal nature.

'Those are fractals at the heart of chaos, my dear. They are the infinity of matter. They are the door, the inscape of the universe,' buzzed Bumble.

'Then my teachers were wrong. Chaos isn't really...' began Alice.

At that moment, Strange Attractor appeared in the mirror. He winked at her again and said, 'You're right, Alice. Order emerges out of chaos and chaos out of order.'

'But how does Chaosland relate to Quantumland. They both seem so different,' Alice asked.

'Chaos is a special kind of order, the infinite order of Newtonville, if you like. It's the inner beauty of the clockwork world. Quantum or classical, whatever way you look at it, the universe is inexhaustible and infinitely varied, like thought itself. So, whenever someone tells you they've found the answer, or discovered the ultimate equation, then they'd better look out, because I'll be just around the corner, waiting to trip them up!'

And with that warning, Strange Attractor vanished. Alice looked up and saw the carousel. Mr. Smith went round and round with his wife. The same pompous-looking passerby walked by...and slipped on a banana peel.

10 QUEEN ROSIE

Nick Herbert

In which Rosie warns Alice about Science's
Patriarchal Bias.

As Alice walked along the Midway, she noticed a mailbox in the distance that had not been there before. On a whim, she walked over to examine the box, when suddenly letters started to pour out in a stream.

'What nonsense is this?' Alice said, reaching down to pick up some of the spilled mail. One was a large brown envelope from the 'Universal Atomic Sweepstakes.' A postage stamp in one corner had a picture of Alice on it.

'That's not right,' thought Alice when suddenly the image winked at her and said quite distinctly, 'Open me please.'

'All right, if you insist,' said Alice, since by now she had become used to strange requests in Quantumland.

Inside, she found a letter that said: DEAR ALICE: YOU MAY ALREADY BE A WINNER!

She found a list of prizes, each with a scratch-off patch next to it.

THOUSANDS OF FREE GIFTS GIVEN AWAY EACH SECOND! TODAY MAY BE YOUR LUCKY DAY! CHOOSE YOUR PRIZE NOW!

1. Sports Car
2. Diamond Ring
3. Vacation in Mexico
4. An audience with the Queen.

'Nobody ever wins these silly things,' thought Alice, while idly scratching at number 3, the vacation in Mexico. The word WINNER sprang from the scratch patch, and she heard an announcer's voice boom out, 'Congratulations, Alice!'

Alice was suddenly enveloped by a shimmering flash of light, which slowly dimmed to reveal a formal English garden with a large rose tree near the entrance. The roses growing on it were white, but there were three Gardeners—card-dressed Spades—busily painting the roses red. Alice thought this to be a curious thing. A most unusual game of croquet was in progress on the velvety green lawn. But the croquet balls were live hedgehogs, and the clubs were live flamingos!

Then with a start, Alice found herself standing next to a rather large woman in Victorian clothes. She shouted out at her, 'Well, don't just stand there like a lump. Pick up a flamingo, dearie.'

By now Alice was very puzzled, 'Where am I? This isn't Mexico!'

'No it's not. You're back in Quantumland, sweetie!'

'Oh, then you must be the Queen,' guessed Alice.

'And yes, in a manner of speaking, I am the Queen. I'm a quantum physicist at Mammoth State. I just won one of those Atomic Sweepstakes, too. So don't be too disappointed you're not in Mexico,' the Queen commiserated. 'I chose dining and dancing on the Riviera with that dreamboat singer Yanni, but instead I got to be Queen-for-a-Day in Quantumland!' she complained, indicating the garden.

'Well, it is a lovely garden in which to spend a day as Queen,' Alice said encouragingly.

'Some Queen! In Quantumland the laws are probabilistic. Half the time my commands are obeyed and half the time they ain't.'

The Queen turned around and commanded. 'Knave, bring this girl a club! Where is my Knave? They're always under foot, but never around when you need one.' She turned her attention back to Alice.

'Oh, well, you can use mine,' said the Queen, handing Alice her flamingo. 'My name is Rosie. You must be Alice.'

'Pleased to meet you, Queen Rosie, but how did you know my name?' asked Alice, still amazed that everyone in quantum reality seemed to know her name and read her thoughts. 'Oh, it must be non-locality!' Alice decided. 'Yes, that must be it...'

'You're still wearing your name tag from Oyster Quadrille,' she replied. 'Skip that Queen business. You can call me Rosie, Alice.'

Alice's amazement was replaced by surprise that Rosie would also know about Oyster Quadrille. 'You know about Oyster Quadrille—the super-secret project to investigate Deep Reality? Do you know Doctor Yes and Doctor No?' asked Alice excitedly.

'Sure, they're both colleagues of mine. A pair of big windbags, if you ask me,' Rosie scoffed.

But they seem so smart and so sure of themselves. And what about Professor Who?' she wondered.

'At least Who's trying to do experiments rather than just talk about Reality,' Rosie grudgingly responded. 'But he's really in the same boat as Yes and No. There's not one chance in a million those bozos will ever discover the secret of Quantum Reality. And do you know why, Alice?'

'No, tell me.'

'It's because they're all MEN!' declared Rosie, triumphantly.

'What's that got to do with it?' queried Alice.

Rosie explained: 'Ever since science began, Alice, it's been a man's game. Or should I say a game for little boys. Little boys attempting to peek up Nature's skirts. Francis Bacon, the father of the experimental method, described his new approach as forcing Nature's secrets from Her by putting Her to the rack. Patriarchal science is not really about understanding Nature, but controlling Her. Power is the name of the game, Alice. Power is what governments are all about, and power-hungry governments pay big money to support certain kinds of science—the kinds of science that promise to give insecure men's clubs more power over Nature.'

'Knowledge is Power, right?' quoted Alice. 'But surely knowledge is much more than power. Don't we gain a deeper understanding of Nature when we learn to control Her? That must be good, mustn't it? And surely the scientific facts themselves don't depend on a scientist being a man or a woman?'

'Every scientist whether she's a man or a woman has bought into the same method. They call it "scientific objectivity." What that means,' said Rosie scornfully, 'is that Nature is treated as an object!—an object to be looked at, to be spied on, to be smashed, to be ogled at by a dispassionate observer who stands apart, unaffected by his observations. They all look at Nature that way; that's how science is taught in the schools. It's true that there's only one set of facts, one science, for both men and women. But this one science happens to be male science; men of both genders standing on street corners leering at foreign objects.'

'And as for understanding,' Rosie continued, warming up to her favorite topic, 'Alice, do you know what Einstein said after fifty years of studying quantum theory? "Who would have guessed," he said, "that we would come to know so much, and understand so little?" He was referring to the fact that physicists use quantum theory as a mathematical tool to predict the results of experiments but have still not yet discovered a consistent world picture—a quantum reality—that explains how and why this tool works. And can you believe it, Alice? Most physicists are content with this situation! Perfectly content to play with quantum power, without seeking quantum understanding.'

Alice was fascinated by these new ideas.

'Do you really believe, Rosie, that there's some way of knowing the world other than the methods of objective observation—what you call "patriarchal science?" What would such a way of knowing look like?'

'Here's the first croquet. I'll let you go first,' said Rosie, abruptly changing the subject back to pitching hedgehogs. Alice was puzzled by the sudden change, but obediently examined the first hoop.

The first hoop consisted of two parallel tunnels going through a mound of earth. On the other side, the tubes became troughs, which first diverged,

then curved together to meet at right angles, crossing one another and loop-
ing back to return to the start. At the crossing point, there was a third
trough that led to the hole.

'Curiouser and curiouser. This is the strangest game of croquet I've ever
seen!' said Alice

Alice chose one tube, and hit the hedgehog, only to have it return to her.
The same thing happened when she chose the second tube. She tried a few
more times, but gave up in frustration.

'What can I do? No matter which tube the hedgehog goes through, it
always comes back!'

Rosie smiled. 'On the first hole, you've got to hit the hedgehog through
both tubes at once. This is Quantumland, after all,' she explained.

A light went on in Alice's brain as she remembered that an electron
could be in two places at once.

'Oh, I understand. Quantumland means wave-when-you-don't-look;
particle-when-you-do-look, right? I suppose that means I've got to hit the
hedgehog without looking.'

'That's right, Alice. You're learning to think in the quantum way,' Rosie
said approvingly. 'Here, let's put on this blindfold.' She took a purple silk
scarf from around her neck, and tied it around Alice's eyes. 'Now try hitting
the hedgehog without looking,' she encouraged.

Alice hit the hedgehog, and this time, it did not return. They looked
over the mound and saw that the hedgehog had settled in the hole.

'Good for you, Alice! Now you blindfold me.'

Alice complied, and Rosie made a hole in one. They moved to retrieve
the hedgehog, but the hedgehog had unrolled itself, and was in the act of
crawling away, so Alice and Rosie walked further into the garden after it.

'One of the lessons quantum theory teaches us has to do with the renunciation of knowledge. The Danish physicist, Niels Bohr, in spite of being male, was very sensitive about the necessity for this renunciation. He
taught that there were certain things in Nature that could never be known,
even in principle.'

'So there's...there is no deep, hidden reality?'

'Yes, Alice, only phenomena are real.'

'So it's like in the old-fashioned horror movies, Rosie?—There are some
things that men should not know'? asked Alice.

Rosie patted her hand. 'Something like that, dear, but worse. In the
quantum world, there are some things that men cannot know. We must
ground our understanding of this world on the certain fact that certain facts
about the world are simply unknowable. And we must come to realize that
this unknowableness at the heart of matter is not merely unavoidable uncertainty but, in some strange and beautiful way, forms the essential foundation of the visible world. I really like Niels Bohr's ideas. I had a crush on
him in college. He reminded me of my father.'

Alice stuck to the main point. 'So we have to give up some knowledge in order to get other knowledge. Is that right? Seems like a fair trade. We still get to keep our objectivity, don't we? What do you think, Rosie?'

'Well, I'm Queen of Quantumland today, but tomorrow I'll just be another dizzy dame. I don't really know all the answers but here's what I imagine to be so.'

'Please tell me, Rosie; I really want to learn.'

'Well, one of the strangest and most wonderful things about quantum theory is that it tells us that the world is totally connected: it's all one big waveform—an inseparable Unity. Yet the very first move we physicists make in our description of Nature is to divide the world into subject and object.'

Alice remembered what the Quantum Hussy told her about how we are waves of possibility on the sea of the Quantum Vacuum, but she didn't see any alternative to Rosie's complaint.

'But how else can we think about the world; how else can we speak about the world?' asked Alice, reasonably. 'Subject/object thinking seems to be built into our brains; it's certainly built into our language.'

'Thinking, that's just it! Too much thinking!' exploded Rosie. 'All men and all females who've swallowed the patriarchal bait put too much value on thinking. Too much head and not enough heart. You know what I say, Alice, as Queen of Quantumland? I say off with their heads, all of 'em!'

'But surely you can't abolish thinking!' said Alice, stunned. 'Thinking's what separates us from animals. Where would we be without it? Why, we'd be back in the jungle.'

'Thinking's fine as far as it goes,' explained Rosie, 'but if we're going to fully understand Nature—including our own nature—we've got to go further, much further than mere reason can take us. I think tomorrow's scientists will consider our so-called "objectivity" to be downright medieval.'

'But where can we go beyond "thinking?" Are you recommending, like those New Age gurus, that we get in touch with our feelings?' asked Alice suspiciously.

'Our feelings are part of it, but we need more,' asserted Rosie, passionately. 'We need many and more complex ways to contact Nature, ways that go deeper than mere "observation," truly scientific methods that call our whole beings into play, not just our intellects. We've got to free Nature—and ourselves—from Bacon's torture rack, and find deeper and more creative ways to interact with so-called objects. We've got to learn to seriously relate to Nature, less like some stranger in the street and more like our own Mother.'

'Stop treating Nature like a collection of objects? Is that what you mean?' asked Alice. 'Learn to really treat Her as Mother? Those are great sounding words, Rosie, but what can we actually do to make all this come about?'

'I don't have any answers, Alice. Only dreams and hunches. My intuition tells me to be alert for ways of merging with Nature, for finding something of myself in Her, at all levels from the grandest phenomena, to the least of Her creatures.'

'Merging with Nature? Are you talking about "empathy?"' offered Alice

'Yes, empathy. Imagine how the world would appear if our sense of empathy was as highly developed as our senses of sight and sound. Imagine what it would be like to enter into the inner life of this world with as much intensity and detail as we are now able to examine its outer surfaces.'

'Do you imagine, Rosie, that the scientists of the future will possess new tools for extending our inner senses? Telescopes and microscopes for magnifying empathy?' asked Alice eagerly.

'I hope so, Alice. But in the meantime, I advise you to keep your eyes open for flaws in the separateness assumption. That's undoubtedly the biggest mistake of patriarchal science. Trust the evidence of your deepest experiences no matter how crazy and illogical they may seem. Try practicing perceiving, remembering and believing in phenomena that ordinary science says are impossible. Why, when I'm doing my best research, I can manage to believe in ten impossible things before breakfast! Look inside yourself, Alice. Trust your deepest instincts for truth.'

Alice thought about what a nice idea it was to think of the universe as an undivided whole, and wondered if maybe Rosie had met the White Queen, who, too, believed in 'impossible' things.

Just then, there were the sounds of commotion and voices in the distance.

'Oh-oh,' she broke off, 'looks like our time is up. Here come the knaves.'

A group of Jacks rushed over and surrounded the Queen, babbling away.

'I'm needed elsewhere in Quantumland, Alice. This place doesn't run itself, you know. And I'm only "Queen for a Day." Remember Alice, off with their heads! Thinking is fine but it only goes part way. Go deeply, deeply, Alice. It's the surest way through. Here's your return ticket. Best of luck, dear.'

Rosie handed her a Universal Atomic Sweepstakes form identical to the one that brought her here. The garden began to dissolve around her, like a chalk drawing washed away by the rain.

'Thank you, Rosie. Thanks for everything,' called Alice.

A moment later, the Midway was coming back into focus. She found herself standing next to the mailbox once more, still holding the sweepstakes form.

'Thank you, Queen Rosie. Thank you a lot,' Alice murmured.

11 ALICE IN IRELAND

F. David Peat

In which Alice meets three outrageous Dubliners:
18th century philosopher and cleric George Berkeley;
19th century mathematician W. R. Hamilton; 20th
century wit, writer and drinker Myles na gCopaleen,
and experiences Quantum Reality, Irish-style.

Alice had become quite confident about her ability to jump between different worlds that she felt she really had no need for the quantum cat to act as her guide. 'It's so simple really. I'll just blink and take myself to the seaside.'

Alice blinked and found herself wandering around the Berkeley campus, content for the moment to enjoy the scenery and inhale the salty San Francisco Bay air.

'Berkeley! Wow!' Alice shouted out loud. 'I really do like this place. What a view, sun, sea, and mountains. No wonder so many people came here. All those quantum wizards, Oppenheimer, Feynman, Bohm, and that's just the start.' She began to hum 'If you're going to San Francisco,' as she happily walked along.

'I remember seeing those movies with beatniks, hippies, flower power, the Mamas and the Papas, Jimi Hendrix, open air concerts. What a place! This is Berkeley!!'

'No, I am, my child.'

Alice spun around. There stood a man dressed in the very elaborate costume of an eighteenth-century bishop.

'What? Who?' Alice asked, stunned.

'I'm Berkeley. George Berkeley to be precise, Bishop of Cloyne in Dublin, Ireland.'

'What a coincidence. You've got the same name. Are you an old hippie? You're certainly dressed in funny clothes.'

'Not at all, my dear child.' He gestured around him. 'All this is named after me—"Westward the course of Empire takes its way, de dum, de dum, de dum, de dum." This was my dream. To establish learning in the New World and extend man's knowledge westward towards the setting sun. Although I'm not quite sure it all turned out like I'd expected! Jimi Hendrix, you said. Would he be a new bishop here?'

'I think you're a little mixed up. And we certainly don't use words like "Empire" or "man" anymore, they're not thought to be very polite. And while we're about it, I'm not your child. My name is Alice.'

'Words, words, words, words. Don't we all owe so much to them. I hope, at least, you drink tar water. I wrote a great treatise on its preparation and imbibing.'

'Tar water?' asked Alice, much surprised.

'I see it remains unknown to you. "Hail vulgar juice of never-ending pine! Cheap as thou art, thy virtues divine..." What a tragedy that tar water should have been overshadowed by my philosophical writings. Here, step into this railway carriage and I'll explain.'

Before Alice could respond, Berkeley pushed her into a phone booth. In the next moment, she found herself with Berkeley and several other companions in a Victorian railway carriage bumping across the lush green Irish countryside.

'Tar water,' Berkeley continued as if nothing had happened, 'especially efficacious in cases of falling hair, blocked bowels, scrofula, housemaid's knee, onset of the nadgers, crossed eyes, decayed teeth, lack of concentration, diseases, disturbances, and disorders of the blood, malfunctioning of the kidneys, lassitude, bucolic plague...'

'Don't you mean bubonic?' interrupted Alice.

'Do I? You do seem to know a great deal, my dear. A pint of tar water a day would bring this country to its feet. Have a glassful.'

'To its knees, he means. Take no notice, Alice, and have a dish of tea. That'll warm your heart,' said one of the men in the carriage.

'A pint of the porter's your only man,' said a second.

'But I never drink,' Alice replied.

'How dehydrating for you!' responded the second man.

'Did someone mention tea?' inquired another traveling alongside the train on a penny-farthing.

'But who are all these people?' asked Alice.

It was Berkeley who answered.

'A few friends, here to enlighten you in your confusion, my child... er, Alice. You've been consorting with some of those rough characters in Quantumland, haven't you?'

'Oh, yes, and it's been so very confusing. But, on the other hand, it's exciting and so new,' she exclaimed.

'New! Poppycock! There's nothing new in all that. We order things far better in Ireland, don't we, Myles?' Berkeley said, turning to the second man.

'It's Brian,' the second man answered.

'I'm sorry. Brian.'

'No, it's Flann,' the second man argued.

Berkeley turned back to Alice.

'Let me explain. Our friend is the proud owner of several names: Myles na gCopaleen, Brian Ó Nualláin and Flann O'Brien for a start, but I prefer to call him the "Wise Man of Sligo" and leave it at that, for he's one of Ireland's wittiest writers.'

'But why do you have so many names?' inquired Alice.

'That's because I write so many books,' Myles answered.

'I've done a lot of things and I only need one name,' said Alice.

'But do you owe any money? And anyway, there's several of you and only one of me,' Myles rejoined. 'Haven't you noticed?'

'That's not true. There's only one of me. I'm Alice, and I'm not at all confused about that, even though lots of people are always trying to convince me otherwise.'

Berkeley spoke up then. 'We'll see, we'll see as the journey progresses. Now take those little green men you've been learning about...'

'The electron's your only man,' Myles interjected.

'I hear that some days you call it a wave and others a particle,' said the Bishop.

'So why not a Brian or a Flann, or even a Myles for that matter?' said Myles.

'But not all at once!' insisted Alice.

'Only when I write myself into my own books.'

'And how do you do that?' She was curious.

Myles explained: 'With the greatest of ease. I once wrote a book about a man who's writing a book about a number of very curious characters who happen to be reading your man's book, or dreaming they're reading it, and all the time he's writing it. Or was it that your man was dreaming he was writing it? I hope I'm making this perfectly clear. So, naturally, they're a bit on edge, as it were, about making too much noise and singing and dancing over a wee drop of the craythur at night and then waking up your man in case he stops what he's doing and they all vanish. Now, can you ease your mind around a way out of that pretty dilemma?'

'Why, no, I don't think I can,' Alice answered.

'Why, but our friends start their own book about a character in bed who's writing a book about them. It all goes around perfectly nicely and everyone can mind their own business. Why, even sometimes Myles is writing about Flann, writing about Brian, who's writing about Myles writing about...'

'Stop, stop, stop!' Alice cried. 'That's impossible. It's like lifting yourself up by your own shoelaces.'

'Or bootstraps,' said Myles.

'Or gaiters,' Berkeley continued. 'And why not? Didn't those quantum friends of yours ever think to tell you where all those electrons come from?'

Alice knew the answer to this one. 'That's easy. Particles collide and produce other particles, like electrons.'

'And what produced the particles that collided in the first place?' quizzed the Bishop.

'Well, er, other particles, I suppose?' guessed Alice.

'Including some of the particles that were produced in the first collision?' he prodded.

'Well, maybe,' she answered, not too sure of herself.

'Particles produce the particles that create themselves.'

'By their own bootstraps. Curiouser and curiouser,' Alice frowned.

Suddenly Alice had a flash of recognition. 'Oh, I know who you are! Yellow Cab told me about you and Bohr and the Copenhagen Interpretation. "All magic wands and soap bubbles,"' he said. 'Am I right?'

'Quite so, young lady. Very good!" Bishop Berkeley confirmed.

'*Esse Est Percipi*' chimed in Myles, 'To be is to be perceived.'

'But let's get back to the thrust of our inquiry,' reminded Berkeley. 'If everything produces everything, then everything is both itself and something else. So why can't a particle have lots of different names, just like our friend Myles?'

'Or instead of lots of different names, like me, there could be one name in lots of different places, like you,' Myles suggested.

'You're making my head spin! None of this is making any sense any more. Things were much simpler in the quantum world. And anyway, I'm not in lots of different places. I'm here. Aren't I?'

'I thought you were in California,' Myles said.

'Oh, do be quiet, or I'll never sort things out!' she snapped.

'Who said it was up to you?' Berkeley asked.

'What?'

'Sorting things out.'

'I bet you can't even sort out the wheels on this train,' Myles challenged Berkeley.

'What wheels?' Berkeley responded.

'Well, there have to be wheels on this train. That's one thing I'm certain of,' Alice declared.

'How do you know if you haven't looked?' countered Berkeley.

'Don't be silly. I'll prove it.' She leaned out of the window and pointed. 'There they are, going round and round.' She sat back in the carriage. 'See, I was right. There are wheels on the train.'

Berkeley smiled and asked again, 'How do you know?'

'Because I saw them.'

'You saw them then, when you looked out of the window. But, you're not looking now. So how do you know they still exist?'

'You can't fool me with that. There still have to be wheels.'

'And for why, if I may make so bold as to inquire?'

'Because...because. Because things just wouldn't make sense any more if they weren't. I mean, the whole thing wouldn't fit together properly—us moving along like this if the wheels only came into existence when I looked at them,' Alice said, exasperated.

'So, "the horse is in the stable and the books are on the table as before!" That's common sense,' Berkeley assessed.

'Thank you.' Alice seemed mollified.

The Bishop continued: 'But common sense is rather common when you come to think about it. It's not very refined and it's certainly not knowledge. Each time you measure the electron, you determine one of its properties, sometimes its speed, sometimes its position. But do you know anything about the electron itself? And what happens when you don't make an observation? What takes place in the intervals between?'

'It must be there!' she huffed.

'Must it?' It was Berkeley again. 'Is the electron no more than the sum of its attributes, its properties, its correlations? And those things are only defined when you measure or observe them. Each time you look you can speak of the electron's attributes—but what of the electron itself? In what sense does it exist, in what sense does it have being? You look and you believe, but what lies in the looking but the sense data of the mind? And where does reality lie, in the thing perceived or in the act of seeing? And you, Alice, what existence do you have apart from my observation of you? If I were but to shut my eyes, would you go on existing?'

'Oh, dear, it seemed so clear before,' wailed Alice with growing agitation.

'Each time you observe one of these attributes you change it. So, in what sense does the electron exist? In what way does it have its being?'

'But there must be something there.' She was beginning to feel a bit desperate.

'Some "thing." But what meaning is there to a thing in the absence of all its attributes? Best stick to tar water, my child.'

All of a sudden another voice said, 'I once thought of an attribute without any thing.'

Alice turned and saw that the man on the penny-farthing was still high-wheeling it alongside the train. It was he who had spoken.

'Oh, it's the man who was asking for tea.' Alice thought he looked like someone else she met once in a dream, but couldn't quite place him. 'I'm afraid that we haven't been properly introduced.'

'Oh, but I have…' he said. 'I introduced you already. But that was some time ago. Dodgson's the name, Charles Lutwidge Dodgson. Let me tell you about the smile. My muse once dreamed up a Cheshire Cat that began to vanish. Bit by bit all its attributes vanished until there was nothing left but its smile; a quality that exists without any referent. What do you think about that?'

'Very nice to meet you, Mr. Dodgson. Have we met before? You look awfully familiar,' Alice asked, then without waiting for a response, continued with the bizarre subject of the conversation. 'I do seem to remember something about a cat, it was in a dream I once had. But a quality that exists without any referent? That's silly. It's impossible. A smile without a cat, that's far worse than the poor cats of…yes it was someone called Herr Schrödinger. I'm really starting to forget everything, the train is making me so sleepy.'

'The smile and the cat are as real as you are. I wrote it down in black and white so we certainly know for a start that it can't be a marmalade smile, or a Persian blue cat, or an orange tabby. And if we know what the smile is not, then it certainly must exist. Just like you.'

'Alice, beware the man on the wily bicycle, or penny-farthing, in this case, particularly if he's English,' Myles cautioned.

'A bicycle can't be wily. It doesn't have a mind of its own, for one thing,' reasoned Alice.

'You're in Ireland now. Let him tell you,' Dodgson responded.

Myles continued: 'Here, bicycles have a mind of their own, or rather a mind of someone else's. That's why they're inclined to go missing outside your house and turn up at the nearest pub.'

'That's silly. A bicycle needs someone to ride it,' she scoffed.

'Which is mainly policemen in these parts,' Myles went on. 'All of which explains why some of that class are inclined to become a little insubstantial from time to time.'

Alice clapped her hands over her ears. 'I wish you'd all go missing right now so that I can stop my head spinning.'

Myles ignored her outburst and persevered with: 'Now, what's a bicycle seat when it's about?'

Alice sighed, 'That's easy. A thing you sit on.'

Myles was triumphant. 'Gong the lady out! It's air, pure insubstantial nothingness like a pay packet by Sunday morning. Oh yes, with a few of those atoms and electrons and things. I'll let you have that. But beyond that class of characters there's nothing much else doing in a bicycle seat by way of entertainment. Now, what do you think of a policemen's bottom?'

'No, thank you very much,' replied Alice firmly.

Myles forged ahead anyway.

'Pure vistas of nothingness, pints and pints of it. Though not quite so empty as what lies within the skull, perhaps. So, what happens when the two of them meet, the bottom and the seat? Lots of space and just a few atoms whizzing about, minding their own business. And who's to say which belongs to a policeman's pants and which to the saddle? Multiply this by all the hours our bobby spends pedal pushing up and down the hills in search of speeding motorists and strayed cattle and you get my drift.'

'You really are a most peculiar man!'

'Ergo, the bicycle becomes a bit like a policeman, with a moderate wish for a little independence on a Saturday night, while your man becomes that little more mechanical the while. Each takes on the other's characteristics which explains the existence of powerful traffic jams on an island without cars and the movement for the emancipation of bicycles.'

'But bicycles can't think,' Alice was proud of this bit of logic.

'There are as many atoms in their seats as in a bobby's brain. And what was it your man, Bohm, was always saying? Electrons have proto-mind. So why not a bicycle that by the hour is becoming more of a policeman? And that's why I always take the precaution to sit on two sheets of folded newspaper—it saves piles of trouble down the line.'

All at once, the men laughed, chortled and guffawed.

Alice was disgusted. 'Oh, leave me alone, all of you. You're very silly people.'

Berkeley finally spoke up again, using his most soothing bishop's voice.

'We want to help you, my child, show you that things are ordered in a far better way here in Ireland.'

'It's simply a matter of having a few more words to game with,' added Myles. 'Given the choice between the next square meal and language, I'll eat my words any day.'

'Or drink them!' said Berkeley with a smile.

Alice squeezed her eyes shut and asserted. 'I'm in America and this is the twenty-first century. You're all nothing but a bad dream. A bishop in gaiters, a man on a penny-farthing, and someone with more names than I've got summer hats. Put a stop to it this minute!'

Suddenly the first man who had spoken to her on the train tried to comfort her. In truth, she had forgotten that he was there.

'Console yourself with a mutton chop,' he said, offering one to Alice.

Alice lost her temper. 'Oh, that's too much. Take that nasty, greasy thing away. It's rude to offer ladies food before they've been properly introduced.'

'Alice, meet William Rowan Hamilton, half of the famous music hall gymnastic balancing act of Hamilton and Jacobi,' Myles offered obligingly.

'Hamilton is a famous mathematician and as fond of a mutton chop as I am of tar water,' Berkeley affirmed.

'First cats and bicycles and now mutton chops?' Alice despaired.

Hamilton carried on, unfazed. 'Think of a piece of fish, by way of illustration. An evasive haddock, your working-class herring, the depressed flounder. What use are they to man?'

'Or a woman!' she retorted, her feminism now ignited.

'I stand corrected,' Hamilton said. 'When they have been consumed, what are they but white, brittle bones? Not fit for toothpicks. Lodge one in your throat and puff, your game's played out like Tim Finnegan at a wake. But a mutton bone serves first as a good breakfast and then a fine bookmark, or if you were to take a pair of them then you'd have bookends. The grease is particularly efficacious on the leather bindings, preserves them I'll be bound. And thus Ireland's humble lamb takes its rightful place amongst the Ivory Tower of learning. It's all a matter of change of use and change of description—call it part of a lamb or a bookend, a wave or a particle. Words have such power, don't they?'

'You certainly don't talk like any mathematician I've ever known,' stated Alice.

'Functions are my life. Mathematics is the bubbling up of pure thought into formality. As someone else once will say, it's the relationship of relationships. And it is a fine sensual joy to go along with it, this dance of the mind. Correlations of correlations, constellations of correlations correlating... Why, when I discovered the function that bears my name, I was walking on the banks of Dublin's Royal Canal at the time. I was so overjoyed that I carved the formula on one of the bridges. And thus an Irishman is the one true originator of graffiti.'

Alice decided to play along. 'Well, with mutton chops for book marks, why not bridges for pages?'

'A very sensible girl, Bishop,' complimented Hamilton. 'Where did you find her? But let me continue. The world around you, what is it? Clouds, trees, stars, planets, rocks and water, all animated in the theater of the mind, ideas forming and reforming, ideas born out of ideas, ideas dying into ideas. The seasons come and go, the decaying tree stump gives birth to new life, rain falls and forms rivers, and rivers turn into ice. Boil the same water and you make steam that drives the train. It's all a dance of movement, a perpetual transformation. But there is an order to that dance. And how are we to describe it, how are we to put into words the underlying patterns of the universe, the hand of the great magician who turns hats into rabbits and water into ice? With what else but mathematics—the generation and transformation of abstract functions?'

'You're taking my breath away,' gasped Alice.

'Mathematics is the mirror of the unfolding orders of the world. Newton thought that it all had to be done in terms of particles and forces.'

'Yes, I know about that, but then quantum theory came along and showed that it was all waves and particles and uncertainty.'

'No, I did,' Hamilton corrected.

Berkeley nodded, saying: 'I told you that things were bettered ordered here in Ireland.'

'The mechanistic view of Newton's is not the thing itself, but only its description and there are many possible descriptions. George here is a bishop, a philosopher, and advocate of tar water; the list goes on. And water is both the liquid rain of the sky, the hard ice of the winter and the steam for the inviting kettle. Yet, under this flux there is an order, something unchanged, not a thing in itself, but a pattern, a form, the score of the dance, an invariant of Nature. And that's exactly the nature of the function I discovered, the Hamiltonian,' he pumped up. 'For, provided that you keep the pattern of the function always the same, you can describe the universe in whatever mathematical language you like. The outer appearances may look quite different, but what counts is the underlying form, the natural order described by my Hamiltonian function. And so I can describe a system of interacting particles as a flux of waves, and waves as a dance of particles.'

'So wave-particle duality existed long before quantum theory?' she asked.

'Indeed it did.'

'Then what's so new, so special about quantum theory?'

Berkeley answered her this time. 'Nothing, really. It was already inherent in my philosophy, and in Hamilton's mathematics. It's in the wordplay of writers, in the geography of the Irish railway, and the transforming power of the bicycle.'

Alice held up her hand. 'Stop! You're going too fast for me again. What have railways to do with this?'

Myles spoke up. 'The problem, Alice, is that we're all wordmen here in Ireland.'

'And women,' she reminded him.

'In the beginning was the...' said Berkeley.

'Words make the thing...' Hamilton said.

Berkeley continued: 'Are we, when all is done and said, are nothing but a web of words?'

'The world's words are popping air,' Myles responded.

'And popping Wiffs!' popped Alice with glee.

'And what are the world's foundations?' Berkeley asked. 'Space, time and matter. Just words, concepts of the mind we place like veils or filters over the body of Nature.'

'They're jelly molds,' Myles chimed in. 'Put the world into different molds and out come your rabbits and castles and mountains and smiling faces.'

'That's where Niels Bohr and his lager-guzzling friends came in with the Copenhagen Interpretation. They said that the words we'd all been using were no longer fit and proper in the New World they had entered. Our words were home grown for our daily commerce with grandfather clocks and bicycles and railway lines, but the poor things simply refuse to go to work at the level of the quantum,' explained Berkeley.

Myles pushed on with, 'Bohr argued that words fix our ideas, words like *causality* and *space* and *time*. They're pages torn from our large-scale world in which everything occupies its place. So what can you expect, he said, when we talk about the atom, but paradox and confusion?'

'All this from a man who never took a drop of tar water in his entire life!' Berkeley remarked. 'You're trying to put the New World into old words, sez he. There's a limit on how far you can go, sez he. A limit is set by words, by language.'

Alice was frowning in concentration. 'Please stop. I'm beginning to remember something. Was it a dream?'

'Of an egg?' queried Dodgson.

'How did you know?' exclaimed Alice, then, 'A talking egg called...'

'That's it!" Dodgson interrupted. 'And he said that words were what he took them to mean. It was a matter of who was master—Humpty Dumpty— or the words! But did we dream it, or was there really a talking egg?'

'Here in Ireland, words have a life of their own and we've learned to live with them,' Myles explained. 'We respect words and, in turn, they let us play with them. And so we watch their meanings flicker. Words are like electrons that pass through a double slit. They complement and interfere with each other, words giving birth to words, words containing the history of the human world, words as deeds and words as dreams.'

'Even our friend Myles, who's a great wordman, lives in the shadow of a great mountain,' Hamilton said slyly.

'Ah, yes, *Finnegan's Wake*, the great book of the world,' Myles sighed. 'That is the book of one man's dream, H. C. Earwicker, that nasty little insect, in which the last sentence is also the first. So, the book's beginning is also its end.'

'Have you thought, Alice, that time may be like that book, curving and twisting and going back on itself? Maybe that's why you've had so many lives,' Dodgson offered.

'You keep saying that, but I'm only in one place at one time and there is definitely only one me,' she stated.

'Take this railway line. It was built by Ham and Shem, two brothers who could never agree,' Berkeley began.

'Another of the world's stories,' Myles smiled.

'And coming to the river that's ahead of us, they quarreled so much that each one built his own bridge, with the result that even today we have to pass over both at once,' Berkeley finished, with the air of a man who has made his point.

Alice didn't understand. 'But that's impossible,' she claimed.

'Possible but uncomfortable,' he corrected.

'And not so bad after a drop of the craythur—it helps you see single!' Myles insisted.

Alice looked out of the carriage and saw a river with two bridges, one orange and one green. In front of the train, the line divided. Back in the carriage, Berkeley, Myles and Hamilton looked like a badly registered color print in a newspaper. Their orange and green identities began to separate and suddenly a carriage, complete with its occupants, flew away to pass over the green bridge while Alice and her orange companions traveled over the orange bridge. Moments later they were reunited.

'Oh, that was a bit disorienting, but very nice to see you again!' Alice exclaimed. 'I think something like that happened when I was in Bohm-land—double slits they called it. Was it very painful splitting up?'

'Didn't feel a thing. How was it for you?' asked Myles.

'It all happened so fast. I'm not sure that I was really in two places at the same time.'

'But what about being in the same place two times?' asked Myles.

'I don't think I understand.' Alice was utterly confused.

'We order things better in Ireland, while you've spent your whole life being dragged along the current of time,' Berkeley stated.

'Well, I can remember having my breakfast and now I'm starting to get hungry again. My hair keeps growing and I can even remember the day I put my dolls away. So time must exist,' Alice insisted.

'But, as with words, who is to be master, you or time? Why drift with the current, why not decide for yourself when you want to be?'

'May I?' Alice asked, excitedly.

As Alice concentrated, everyone's gestures reversed and the carriage went backwards over the bridge.

'Now, I'll go forward,' she said, and concentrated again. Alice moved forward in time and traveled with the green carriage.

Full of wonder, she burst out: 'I'm beginning to get the idea. I did go over both bridges, but in different times.'

Berkeley smiled and nodded. 'Now you see it.'

'But if that's possible, and I keep jumping back and forward in time, then I could be almost everywhere at once.'

'You're almost there, Alice,' Dodgson urged, panting and peddling along.

Alice pushed on, thinking aloud. 'And maybe the electron's like that. Maybe it's everywhere at once. But wait...that would mean... That would mean that there are not millions and millions of electrons in the universe but just one—a sort of quantum time traveler who jumps back and forth on the track of time to appear everywhere!'

'And sometimes it even meets itself on the stairs coming home at night,' Dodgson responded.

'Curiouser and curiouser,' said Alice, full of awe.

'Are you beginning to recognize me?' he asked.

'I don't think I like this,' she said, feeling edgy.

'It's the name, Alice. Like me he sometimes masquerades,' explained Myles.

'Charles Lutwidge Dodgson, otherwise known as Lewis Carroll,' agreed Dodgson.

'*Alice in Wonderland!*' Alice cried, finally understanding.

'Now, you are beginning to see how you can be one Alice in many different places, as am I. Carroll introduced you first in a children's book I wrote down, but then you took on a life of your own as many others took you and set you in their own locations, situations and thoughtforms,' Dodgson offered.

'But wait, I'm real, I'm me, I'm Alice.'

'One Alice in many different times,' he said.

'You mean I'm the Alice of *Alice in Wonderland*? But that couldn't be. Could it?' Alice paused, trying to realize wisps of impressions and memories coherently. 'No, I know you're not real. All of you, this carriage, Ireland, railway lines that separate are part of my dream. And since I'm dreaming you, how could you have created me?' she demanded to know.

Myles threw up his hands. 'Here we go again.'

'I'm Alice and I'm in Berkeley,' she stated firmly.

The carriage dissolved and Alice was back on the Berkeley campus.

'There, I was right. I'm back in the twenty-first century with cars and television and Internet and virtual reality and quantum theory. The good, old familiar world again. And if there are any paradoxes then they only exist in my dreams.'

'Do they?' a voice said.

Alice looked up. In a tree nearby sat that familiar feline smile without the cat.

And the smile said: 'The question is, Alice: 'Who's minding the store?'

APPENDIX

Flann O'Brien (1911-1966) was one of the names used by the Irish writer Brian Ó Nualláin who also wrote for the *Irish Times* as Myles na gCopaleen. His comic novel *The Third Policeman* explores the relationship between bicycle seats and the bottoms of fat policemen.

George Berkeley (1685-1753) was Bishop of Cloyne and an Empiricist philosopher who held that things only exist by virtue of their being perceived by the senses. He also looked to the New World as offering hope for the future, 'Westward the course of empire takes its way,' and proposed building a college in Bermuda for the native inhabitants. His book *Siris* of

1744 was both a treatise on the virtues of tar water *and* a discourse on the great chain of being!

William Rowan Hamilton (1805-1865) was a Dublin mathematician and astronomer who developed a theory known as quaternions whose full significance was not appreciated until the development of quantum theory—indeed one of the key functions in the theory is known as the Hamiltonian. In later life his diet consisted of alcohol and mutton chops—after his death a large number of bones were found sandwiched between his papers.

12 THE LOOKING GLASS OF ART

John Briggs

In which Alice experiences the Self-Reflexive
Reality of the Artist.

'Well, dear cat, that was quite an adventure, and what strange people
they have in Ireland,' Alice remarked. 'But now I'm back by the sea so I
think I'll stay here for a few days and not make any more trips. And to
prove it I'm not going to blink, ever.'

The smile looked back at Alice. Then a single eye appeared above the
smile. That eye blinked once and the city of Berkeley vanished and Alice
found herself in a very strange Victorian house.

'Is anyone here,' she called out then wandered down the hallway into
a dark, friendly parlor. Dusty light beams splayed through the gaps in half-
drawn curtains. Two walls towered with books. On another wall, behind a
piano, glowed a chaotic collection of paintings. A fire snapped and sighed in
the hearth. A fat cat dozed in front of a mirror on the mantle over the fire.
The cat didn't belong to Alice but it might well have done, because when it
stood up, it arched its back, stretched out its paws and padded off through
the glass. Alice climbed onto the mantle and followed it.

'I seem to be doing odd things all over again,' she thought to herself as
first her hand, then her arm and then her whole body melted through the
mirror, causing her to slide rather precariously off the mantle into a pool of
otherness. She landed with a thud against the Otherside hearth.

'Goodness, don't hurt yourself,' she heard a reassuringly familiar voice.
She immediately recognized the young man bending over her. It was the
man on the penny-farthing who had ridden alongside the train. The man
who said he had written her down into a book. Yes, she began to remember. His name was the Reverend Dodgson, the alter ego of Lewis Carroll!
Suddenly memories flooded in. He'd been so shy as he'd entranced her and

her sisters with the Wonderland stories while on their rowing expeditions up the river Thames. It felt as if she hadn't seen him for a century or more, but that couldn't be right because they'd just been together in Ireland, or so it seemed.

'I'm fine, so far I never really got hurt on these adventures,' Alice assured him, 'just confused. Where in the dickens have I landed now, can you tell me?'

She stood up and looked around her. On closer view, this room on the Otherside looked entirely different from the parlor reflected in the looking glass above the mantle. Instead of containing stuffed chairs, library books, paintings and a piano, the space was entirely filled with mirrors. Small baroque mirrors, large French Empire mirrors, round Chinese mirrors, mirrors in the shapes of globes and triangles, mirrors stacked upon mirrors, mirrors suspended from mirrors. Even the room's walls were mirrors. So many mirrors she couldn't tell the room's size or shape. But the most curious thing was that each of the mirrors seemed to be reflecting an object or scene that was nowhere in sight! In one small hand mirror, hanging by a cord from the frame of a full-length mirror, was a miniature and somehow distorted reflection of the scene in the interior parlor where she'd begun this new adventure.

'And where's the cat?'

'Oh, the cat's from another story,' her friend Rev. Dodgson assured her. 'I'm sure he'll turn up. Let me introduce you to my colleague here.'

It was at that point that she realized there was another figure in the room, a thin, tall fellow with a mustache.

'This is Professor Sggirb.'

'Sggirb? It sounds like something that comes out of you when you slip on the ice.'

'Yes, quite, in a way slipping is Professor Sggirb's specialty,' said Dodgson. 'You see, Alice, my dear, you've been hearing from scientists, and of course I'm a mathematician, so I've asked Professor Sggirb to talk to us about art. He advises me that it's a slippery subject.'

'Slippery?' Alice wondered, 'I hope you don't mean like a worm.'

The Professor cleared his throat. He had a low voice, barely audible, and Alice had to strain to piece together what he was saying across gaps of words she couldn't quite make out.

'Not exactly like a worm, more like the slip of the tongue or the slip of a thought, or a lost key that has just slipped out of reach,' Professor Sggirb explained. 'I'm sure you know what I mean. You remember the difficulties you had with the "Jabberwocky" poem the White Knight recited and the one Tweedledee sang, "The Walrus and the Carpenter," and the one your friend Rev. Dodgson mused that you liked so much, "The Hunting of the Snark." Just to name three.'

'It's true, I was never totally happy with the explanations I heard about those poems. There always seemed a lot more to them—or less,' Alice said, thinking about all the strange creatures who had given her poetry instruction over the years.

'That's what I mean by art being slippery. Rev. Dodgson here has asked me to show you how that slipperiness might be related to the quantum slipperiness people have been telling you about lately.'

'I did think Humpty Dumpty's explanation of "Twas brillig, and the slithy toves" was illuminating, however,' Alice finished her thought.

'Perhaps we should sit down,' Rev. Dodgson suggested.

'Good idea. How about over here,' Professor Sggirb indicated, turning to his right and plunging headlong through the surface of a full-length mirror. As the professor walked away on the Otherside, Alice noticed that the mirror contained the reflection of a business establishment on a London street. The sign over the door said 'Scrooge and Marley.' She and Rev. Dodgson followed.

The cold, fresh air fluctuated with snow falling onto the street. Professor Sggirb pulled a skeleton key from his pocket and opened the door. Inside the old office, they brushed snow off their clothes.

'Don't worry,' Sggirb said. 'Bob Cratchit's not toiling away at the moment. I'm sure you recognize where we are.'

'It's Ebenezer Scrooge's office from Mr. Dickens' *A Christmas Carol*, one of my favorite stories,' Alice identified it with delight.

Rev. Dodgson pulled out some chairs, fetched the scuttle and began stoking the coal fire in the grate.

'Cranky old Mr. Scrooge won't like that; coal's expensive,' Alice cautioned.

'What Scrooge doesn't know about won't hurt him,' Rev. Dodgson replied, then laughed, 'Or perhaps it will. Wasn't that what the ghosts of Christmases tried to tell him? Anyway, whether Scrooge likes it or not, I have Mr. Dickens' permission as a fellow author to utilize these premises. He owns the building.'

'Well, I do think it's rather daring of you, for such an awfully careful man,' Alice teased. Dodgson blushed.

Meanwhile, Professor Sggirb had seated himself on the tall stool behind the tall clerk's desk, presumably Bob Cratchit's. Alice decided that learning about art was rather fun so far, but she couldn't imagine how it fitted into all these scientific peculiarities she'd been learning about lately.

'I was just getting to that,' Professor Sggirb said, reading her mind. 'Remember your visit to Professor Bohm's universe?'

'I certainly do. That Yellow Cab gave me a hair-raising cab ride through New York City. At least I thought it was New York. But, then again, are we really in London now? When you add your imagination to the picture, geography becomes such a muddle.'

Sggirb continued and Alice had to lean forward in her chair to catch his words.

'Professor Bohm was a very brilliant and subtle fellow. He invented several different ways to talk about the quantum world. One of them was the "Quantum Potential" idea you heard about on your cab ride. Let me tell you about another. As you know, in Professor Bohm's universe everything is a part of everything, "enfolded" in everything. One of the images he used to explain how the enfoldment works is the hologram.'

'Yes, I remember seeing a film with a Yellow Cab in it, and yes he talked about holograms but he didn't really explain what they were.'

'It so happens I have a hologram with me.'

Sggirb climbed down from the stool and opened his rather moldy, leather portmanteau that Alice noticed he'd been carrying since they left the room full of mirrors. He rummaged around and held up a piece of glass. It looked like the surface of a pond where someone had thrown a handful of pebbles, causing overlapping rings bumping into each other.

'There's a picture here,' Sggirb explained.

'You must have better eyesight than I. All I can see are rings,' said Alice.

'Wait a moment,' Sggirb propped the edge of the glass against Cratchit's inkwell and rigged up what he called 'my futuristic battery-powered laser.' Then he clicked on his laser, using it to aim a beam of light through the glass plate. Magically, a three-dimensional image of the White Rabbit peering at his pocket watch appeared in Rev. Dodgson's lap.

'You see, the plate with the rings on it records an interference pattern of light. That pattern enfolds information about what the White Rabbit looked like when a holographic surveillance camera took his picture. Now here's the interesting part.'

Sggirb wrapped the glass plate in a cloth, took a hammer out of his portmanteau, and tapped the glass, breaking it into several pieces. Then he mounted one of the plate's fragments against the inkwell and shined the laser beam through it. The image of the White Rabbit appeared again in Rev. Dodgson's lap, though a bit dimmer and from a slightly different angle.

'You see,' Sggirb explained. 'It turns out that information about the whole image of Mr. Rabbit is enfolded on every part of the plate. So in a hologram each part contains the whole. Professor Bohm showed that each quantum particle also contains an image of the whole universe. Remember that in Quantumland a particle is also a wave. So in Professor Bohm's view each particle is an interference pattern containing a record of all the other particles it has met in its entire lifetime, and also all the particles those particles met in their lifetimes.'

'That's a lot of information,' Alice concluded. 'That's even more complicated than all these adventures Rev. Dodgson invents for me to get into.' Alice shot the Reverend a look that once again made him blush.

'Remember, also,' Sggirb persisted, 'that you're totally made up of quantum particles, so you contain all those patterns of information, too, in billions and billions of varieties.'

'So I'm like that room full of mirrors?'

'Yes, in a way. There's a very old saying from one of India's sacred books, *The Flower Garland Sutra*, which is very much like Professor Bohm's idea. It says, "In the heaven of Indra, there is said to be a network of pearls, so arranged that if you look at one you see all the others reflected in it. In the same way each object in the world is not merely itself but involves every other object, and in fact is every other object."'

'That last bit sounds a little like chaos and fractals I learned about in Chaosland.'

'Excellent,' Sggirb said, 'Now you've read my mind. What do you know about fractals?'

'Weelll.' Alice wrinkled her brow trying to remember what that dynamical fellow named Strange Attractor had shown her. 'I remember that chaos is an idea about how things go in the large-scale world we live in, while the quantum is about the very small world we can't see. I seem to remember that chaos means systems, such as the weather and waves in the ocean, that are ghastly complicated and unpredictable, yet follow delicately describable loops and swirls. When weather and waves work away at a coastline for a long time, don't they crumble it up to make a fractal? And isn't a fractal where the small parts look like the bigger parts and also like the whole thing?'

'Splendid. I'd say you remembered quite a bit. Nature is full of fractals where the small mirrors the large. The shape of a tree is mirrored in its twigs; the shapes of the bronchial tubes and air sacs in your lungs are similar at different scales; even the billows of clouds look similar on the small scales as they do on the large ones. That's because in systems on the edge of chaos—and those are almost all the interesting systems in Nature—everything is connected by feedback to everything else. The feedback insures that things in Nature show a "self-similarity." So all living things from ants to antelopes have a similarity, even though in many other respects they're different.'

'Like the interference pattern of each photon is both different and similar and they're all connected to each other.'

'Right.'

'I don't mean to be rude, professor. But what does this have to do with art?'

Rev. Dodgson laughed conspiratorially at the impertinence of his creation.

'I'll show you,' Sggirb said. 'Follow me.'

He made his way into Scrooge's inner office and opened the door of a tall wooden locker usually filled with counting-house forms, but in this circumstance containing—the reader has probably guessed it—a mirror. Professor Sggirb stuck his arm straight into the mirror and followed it with his right leg. In an instant he was swallowed up in the pool of silver. Alice and Rev. Dodgson followed behind. Alice felt a peculiar buzz as she passed through.

On the other side of Scrooge's mirror was what appeared to be a vast museum containing every conceivable variety and type of art: Egyptian sphinxes, Impressionist landscapes, Japanese bird paintings, portraits, Neolithic cave drawings of bison, and all manner of photographs. She recognized some of the paintings as one's she had seen in the dark, friendly parlor where this latest adventure began.

'Actually, we're now in a virtual museum,' Sggirb explained. 'We've stepped into cyberspace. I won't bother to explain what that is. But for our purposes it means that replicas of all these artworks can be brought together in the same place and time. The important thing for us is the artworks themselves. That's what we've come to look at. We could examine any of them and find some form of what I want to show you, but let's stop at this one.'

They gathered in front of an old paper scroll about three feet long and fifteen inches wide with a figure drawn in ink. The title under it said *Man with Umbrella*, after Kao Ch'i-p'ei (AD 1672?-1734).

'I see an old Chinese man standing on a bridge holding an umbrella. What am I supposed to see?' Alice asked, quite puzzled.

'The fractal and holographic nature of this drawing.'

Alice stared at the scroll for a minute but didn't see anything particularly chaotic or quantum in its nature. Though it was a nice drawing.

'What we're looking at in a work of art is not exactly the same kind of order we find in fractals and holograms, but it's similar. I call it the order of "reflectaphors."'

'Reflectaphors. That word sounds strange, but I feel like I recognize it.'

'It's a portmanteau word,' Rev. Dodgson exclaimed.

'Right,' said Sggirb. 'It's a word made by blending pieces of other words together. This one is a rather obvious and academical portmanteau compared to yours, Reverend, which are wonderfully poetic. My word's just made out of combining the words "reflection" and "metaphor." But maybe it will help us understand something. Alice, how would you describe the umbrella the Chinese man on the bridge is holding?'

'Humm. I suppose I'd say it's a triangle with a little top knot.'

'Good. Now, if I were to tell you that there are other shapes like this in the painting, could you find them? You too, Reverend. But I suggest to you that this isn't a game. The other versions of the shape are subtle and they work together to give the painting a mysterious movement and depth.'

They looked at the drawing a while and then Alice said, 'Is that one, the hat on the man's back? It's a bit like a soggy triangle with a topknot.'

'Goodness,' said Rev. Dodgson, 'so it is. And if you trace out these rather darker lines from the man's shoulder, down his back and right leg and up to his armpit you have most of another soggy triangle. The little ink spot on the buttocks is a variation of the topknot.'

'And there are triangles in his trouser legs,' observed Alice.

'And there's a little one under his armpit,' Rev. Dodgson pointed out.

They went on for several minutes discovering other implicit variations of the umbrella shape. It was Rev. Dodgson, the mathematician, who noticed that the drawing seemed to contain a tension between Euclidean triangles represented by the umbrella and Riemannian triangles. He mentioned his discovery to the other two.

'What's a Riemannian triangle?' asked Alice.

'Imagine trying to draw a triangle of the surface of the sphere. All the sides of your triangle would be rounded. This kind of triangle was invented by a Georg Riemann, the great German mathematician who discovered non-Euclidian geometry. The hat looks like a kind of Riemannian triangle.'

'Oh, then look at the bridge,' Alice said, 'If we take Rev. Dodgson seriously—but I should never take him too seriously!—the top edge is Euclidean and the bottom edge is Riemannian, and the man crossing it is like a big topknot. My, this is a fun game, a great puzzle!'

Sggirb harumphed and his voice was more audible than usual. 'Sorry, as I said, it's not a game. I only wanted to suggest there's an order going on here that looks something like the order of holographic quanta and fractals. But reflectaphors go much deeper.'

'I know you said a reflectaphor was a portmanteau word combining the words reflection and metaphor. But just what *are* reflectaphors?' Alice pressed him.

'In a conventional metaphor two things are compared which you ordinarily wouldn't think of joining or juxtaposing. The two things reflect each other but they also go very strangely beyond the reflection because they're also unlike each other.'

'Such as my portmanteau word, "slithy" in "slithy toves," of Jabberwocky fame,' the Reverend put in.

'Exactly,' the professor said. 'If I remember Humpty Dumpty's explanation correctly, slithy is a combination of lithe and slimy. Lithe, Dumpty explained, means the same as active. But of course there's much more to it. Slithy sounds like a lot of things. Its meaning is poetic, slippery. You almost *know* what it means, *feel* what it means, but you can't pin it down. It seems quite vivid and clear, but it's a mystery, too. So it gives you a little jolt like overturning a rock and finding a bright blue and yellow worm underneath. The purpose of every literary metaphor is the creation of life, or something very much like blue and yellow worms, in all their precise vividness and un-know-able-ness.'

'Yes, that's it' the Reverend agreed.

'Puns, portmanteau words, irony, similes, literary characters—all of these juxtapose things which reflect each other in a way that challenges how our mind organizes the world. They inform us that reality is far richer and more creative than our explanations could ever tell us.

Now, I'd like to recite a reflectaphoric image made by Wallace Stevens, an American poet. He wrote these lines many years after Rev. Dodgson and Alice Liddell, the little girl who was the model for you, were no longer alive.'

'I'm very sorry to hear that Rev. Dodgson is no longer alive. I suppose that means now he's imaginary like me,' Alice grinned, impishly. 'Which means we can be anything we want now, doesn't it?'

'Before you two get carried away, let me give you the lines,' Sggirb intervened. The two leaned forward to listen attentively, but now the professor's voice was loud knocking them back.

Among twenty snowy mountains
The only moving thing
Was the eye of the blackbird.

'Now once again, but this time, close your eyes,' prompted the professor. He repeated his performance.

'Why, that's just breathtaking,' said Alice, 'and I can certainly see the contrasts in my mind's eye. Those mountains are huge and solid and white and unmoving and there are twenty of them. The blackbird's eye is small, quick moving, really black, and it's the only thing living.'

'And?'

'Well, it does seem like something's going on in the picture,' Dodgson added. 'But I can't tell whether it's something marvelous or ominous or just everyday. There's just a strong impression that something's implied, but I can't say what it is.'

'So there's a feeling of vivid meaning that's impossible to paraphrase?' asked Sggirb.

'It's a kind of is-ness in the picture,' Alice felt.

'That's it. A great novelist, Virginia Woolf, said that in her work she was trying to describe "moments of being."'

'And being has the mysterious order to it,' the Reverend Dodgson assured them.

'But so does nonbeing, according to the Buddhists,' said Sggirb, 'And maybe being and nonbeing, and emptiness and fullness, are like the high contrast between the eye of the blackbird and the twenty snowy mountains. They're both incredibly different and somehow suggest a sameness.'

'But what about all those different kinds of triangles in the Chinese drawing?' asked Alice.

'I think they are there because the mind is tuning into that mysterious order, the reflectaphoric order, in which different things are the same.'

'Whose mind, professor?' Dodgson asked whimsically.

'The mind of the artist, first, and then our minds as we view the drawing.' He paused for a moment.

'Now consider this. You're both familiar with Beethoven's Fifth symphony, right? Remember how the first measures go: Da Da Da Daaa, Da Da Da Daaa (lower). Beethoven establishes a pattern in the first measure and then juxtaposes it to the slight variation in the second measure. Then he keeps juxtaposing variations. Each time our mind expects to hear the pattern again, and it does up to a point, but we're also continually surprised because of the way the pattern is varied. Even if you've heard the piece many times before you're surprised. Even if you've read the blackbird lines many times before, the reflectaphoric contrasts and the implied similarity keep them moving.'

'So the triangle patterns are reflectaphors that are part of what's making the man on the bridge seem alive?' Alice asked.

'Yes, it's because of what happens to our minds in the gap.'

'What gap, professor?' Alice wanted to know.

'When you reflectaphorically juxtapose one thing to another, we see the two things are related, but we can't fit the connection into our familiar scheme of things. The fact that the items are both simultaneously similar and different creates a gap. The gaps between the terms or contrasts of reflectaphors put us in touch with that unspeakable order of being—or perhaps it's nonbeing—the order that permeates everything, but is always deeper than what you can say about it.'

'It's the blackbird's eye in the snowy landscape?' Alice offered shyly.

'Yes. Juxtaposing that eye to the twenty snowy mountains creates the gap. They refresh us. That's what reflectaphors do. They make gaps. Or maybe they reveal the gaps that exist all over the place in our knowledge about things. Art helps us notice them. But reflectaphors don't always have to have two or more explicit terms to make the gap.'

Alice looked puzzled.

'Look at our man on the bridge. How would you describe the man's facial expression and his attitude as he's crossing the bridge?'

'I'd say he looks like the wind is blowing and he's hunched over against it,' Alice surmised.

'Maybe he's bent over because he's tired. He looks like a determined fellow,' the Reverend said.

'I think he looks a bit sour.'

'Well, I can see that, Alice,' the Reverend said, 'but he also seems like he's thinking about something he has ahead of him.'

'Or something he's coming from,' Alice said, reversing her position.

'And look at all that space on the scroll above him,' Dodgson pointed out. 'He looks like one lone individual moving in all the vast space of the universe.'

'But I think he looks like he's going off to meet people,' Alice countered.

'Actually he looks like me after a long day of reading my students' mathematics papers!'

'Or me trying to get through some of these madcap adventures!'

Sggirb interrupted them. 'You see? There are lots of things we could say about the man, and some of them would be logically contradictory. But they wouldn't be contradictory in terms of the picture. Each thing we could say would touch on some aspect, but nothing we could say would ever come close to saying who—or what—this man on the bridge is. He's a one-term reflectaphor. He's like something, or is he something unstated? Maybe we could even conclude that in some subtle way he's like—or he is—everything. He's a mirror. What's he reflecting?'

'I don't know,' Alice said.

'Precisely. He's a reflection of X,' the professor proclaimed with triumphant obscurity. 'Between him and what he's reflecting lies the reflectaphoric gap, the unknown, where the truth is.'

'Professor slipperiness making us slip up again,' Rev. Dodgson laughed.

Sggirb muttered, 'But maybe a better word for slipperiness is the unknown. The man crossing the bridge exits at the junction between the known and the unknown. A poet friend of mine once said that poems made out of metaphors, ironies, portmanteau words and other reflectaphors are really new words for things we don't have words for. And all those new things are really reflections of the same thing.'

'I would call that same thing the magic of being alive,' Rev. Dodgson said.

'Perhaps. Poems, paintings, music—each one makes reflectaphors using different materials in different ways in order to encounter that Whatever-It-Is, X,' said the Professor.

'Well, if you ask me, I'd call it finding out the world is a dream,' Alice remarked.

'Perhaps Alice agrees with Tweedledum's argument,' said Dodgson.

'What do you mean?' Alice couldn't remember.

'Don't you remember? You said to Tweedledum, "If I'm only a sort of thing in his dream, what are you, I should like to know?" "Ditto," said Tweedledum. "Ditto, ditto!" cried Tweedledee.'

'Slippery, slippery,' one of the three in the museum said, it wasn't clear which one, perhaps all three, for the sound reverberated so.

'So Ebenezer Scrooge, Captain Ahab, Hamlet and Jane Eyre all found out the world is not just our ideas. It's not just what we think is separating and joining us. The world is Big Reflecting X. That's clear.'

'But the Big Reflecting X is everywhere if you look. Like William Blake said, isn't the universe in a grain of sand?'

'Or the blackbird's eye against twenty snowy mountains.'

'Or a man with an umbrella crossing a bridge.'

'Or the Mad Hatter's tea party.'

Alice realized that all the while they'd been talking they were walking past wall after wall of paintings, turning down corridors, a veritable labyrinth, sometimes passing the same paintings again but now they appeared to be in different sizes. She was thoroughly disoriented but stayed close to Professor Sggirb. All of a sudden he took an abrupt turn—possibly left, possibly right, maybe both—and they passed through a large set of glass doors—mirrors, naturally—into an Oriental garden.

Here were lovely quiet pools, the gurgling of water, rocks like claws growing out of the earth, gravel along the stepping stones raked up in waves, trees that made her feel in the middle of a windswept landscape except everything stood perfectly still. The arrangement of objects in the garden seemed to reflect the movement of the planets and yet it was made of simple and perfectly ordinary things of Earth. It was an uncanny mixture of the man-made and the natural.

'So you see, through their countless varieties of reflectaphors artists discovered the holographic, fractal universe long before scientists did,' Sggirb said as if carrying on the previous conversation, 'and they discovered something deeper than pattern. They discovered what you can't discover.'

'But how could they discover what you can't discover?' Alice couldn't fathom just what Sggirb meant.

'They discovered it and undiscovered it and discovered it again all at the same time. That's what reflectaphors do. That's why art is so rich and slippery. That's why writers pluck you—Alice—out of Rev. Dodgson's, er, Lewis Carroll's books to explain new and subtle ideas that have arisen in science. Even as we speak you're here, as slippery as a Jabberwocky helping to explain what can't be explained.

Then Alice remembered what that zany Professor Flow warned her about physicists being unable to agree. His words echoed in her ears: 'But I warn you, getting any two quantum physicists to agree on a description of Quantum Reality is more difficult than getting a Jabberwocky to describe a Snark!'

'So it's because I'm as slippery as a Jabberwocky,' Alice wondered, 'that's why I'm falling into a book about quantum physics.

'Mind your feet.'

She almost slipped on a root. Without her realizing it, they had passed out of the garden into an actual forest. She began noticing now all the self-similar and reflectaphoric forms around her, each wonderfully different from her but reflecting her in some way. How would she describe it? The gall on the huge oak tree was like rough bubbles, like a thought bubbling up in her own mind; the spaces between the boulders were like the feeling of misunderstandings; the fallen twigs and branches were like forgotten memories.

As they crunched through the undergrowth, beside the twist of a stream she saw a shining plant. It looked very odd and dreamlike in the forest setting. On closer examination she saw it was a plant with leaves of different sizes radiating in all directions, in each one a different Alice reflected.

Alice mentally stomped her foot.

'So now I see,' Rev. Dodgson continued, 'that you must go on with your adventures because you're a reflectaphor and others will claim you.'

This statement produced in her the oddest of feelings, as if she didn't exist except as the relationship of looking at the flower. She hadn't even felt herself disappear. Well, the truth was she hadn't disappeared.

'Perhaps we all are,' she heard the Reverend's voice saying somewhere behind her as she felt herself lifted up into the forest, spinning out in a thousand shimmering reflections, journeying through untold minds. She was still holding the plant. In one of its curious leaves was the cat climbing down from the limb of a tree.

13 BIOLOGYLAND

Amit Goswami and Nick Herbert

In which Alice is exposed to Controversies concerning the Meaning of Life.

'Well, Mister Quantum Cat,' said Alice, 'I really do think I've had far too many adventures for one day.'

But the cat shook its head, 'Oh, my no, Alice, there's far too much to experience. Look in here!'

He pulled aside a branch and Alice saw a street with what looked like the booths of a country fair. And there was a very large sign: WELCOME TO BIOLOGYLAND.

'Ah, yes,' thought Alice, 'so this is a land dedicated to the understanding of living things! Maybe I can find the answer to the question of who I am in Biologyland?'

She began looking around with great anticipation until she heard a very authoritarian voice right behind her.

'You seem new here.' It was a serious-looking man in uniform who addressed her. 'Here in Biologyland, all newcomers must learn the basic doctrines of biology before proceeding any further. Into one of those booths, now!'

Alice was miffed. Shouldn't the sponsors of Biologyland offer a little hospitality before they imposed their doctrines on their guests. But then she remembered that even her own country required newcomers to submit to immigration procedures before being allowed in. So, deciding to conform to their customs, she walked through the door the officious man had indicated.

Inside the booth, she saw charts, pictures, and push buttons that switched on videos with animated explanations of various theories. There was quite a crowd studying them. Basic biology, Alice quickly learned, is composed of three things: a doctrine called *Neo-Darwinism*, a *Central Dogma*, and another doctrine called *Genetic Determinism*.

Neo-Darwinism, a professor on the screen was telling her, is the idea that all life evolved from a common origin, a single living cell, over hundreds of millions of years by a doubled-barreled mechanism.

The first barrel of this Theory, called *Chance Mutation*, produces variations in the genes, the basic hereditary components of all species. Genes are arranged on the DNA molecules inside the cells that make up all organisms. DNA was spelled out in full, too, 'deoxyribonucleic acid,' but that was too much of a mouthful for Alice to even think about.

The second barrel of Neo-Darwinism is called *Natural Selection*, in which Nature selects among the variations offered by gene mutations the ones that will assure the survival of the species. Chance mutation and survival necessity are the sum and substance of biological evolution in Neo-Darwinism.

Simple enough, thought Alice, *but just who is doing the choosing?* She held that thought back and continued on.

Through variation and selection Nature produces dynamic alterations in the characteristics of a species that enable it to adapt to environmental changes in order to survive. But when the accumulated alterations are so radical that interbreeding with the pre-existing species becomes impossible, then we say that an evolutionary jump has occurred from the old species to the new one. *And who makes the decision for that to happen?* Alice had questions, but knew she would have to suspend her disbelief to get benefit from the exhibits.

As evidence of evolution, the charts presented enormous volumes of fossil data. Alice perused them, finding the evidence of the creation of new species and of extinctions, all quite compelling.

So all species were not created at the same time. More complex species usually appeared later in time, which seemed to be saying that evolution is something about going from the simple to the more complex. There was also fossil evidence of enormous adaptive changes within a species, taking place over millions of years. *Very impressive.*

But gradually Alice noticed something was missing. In all this massive fossil evidence, there was not one sequence that clearly showed the intermediate evolutionary stages leading from one species to another. True, such intermediary species might not have been fit to survive for long, but they had to survive long enough to make the link. And the fossil data should show remnants of them, somewhere, sometime.

'But where are they, where are the fossils of the missing links?' Alice asked herself aloud.

'Shhh,' somebody said from behind her. It was a young man of about sixteen who smiled when she looked at him. But Alice was too preoccupied with her thoughts to think much about him.

'So, Neo-Darwinism says that I am descended from the apes through chance mutations and the survival necessity. That sounds more like those nasty predators and banksters who I understand live on Wall Street! Why, that's not me. I know there's more to my creation than that!' Alice murmured under her breath. 'But wait, maybe the next two doctrines will help me understand.'

Central Dogma seemed simple enough. Information flowed one way, from the DNA to the protein and never the other way around; that is, never from the protein to the DNA. Alice was wondering what was so deep about that when somebody on the video explained its meaning: environment may directly change the proteins of an organism, but such changes are not passed on to the DNA or the genes, and thus are not reflected in heredity. In other words, development of organisms has no effect on their evolution.

If I understand correctly, Central Dogma says that DNA does not learn anything so nothing new gets created and passed on to the next generation. That doesn't sound like evolution to me, Alice thought.

Genetic Determinism, Alice discovered, was the idea that everything about a species was determined by its genetic composition. 'That's interesting! I am nothing but my genes,' Alice chuckled when she encountered this strange notion: 'A human being is a gene's way of making more genes.'

'Yes, indeed!' the detailed explications of the doctrines told her, 'You are a gene machine.' This was repeated a number of times in her video lessons. 'Your form, your intelligence, even your consciousness, all are survival tactics for your genes. There is no other necessity but survival, survival of the genes.'

'But I love my cat and surely my love has nothing to do with my survival.'

She must have spoken aloud again, because there was that 'Shhh' once again. It was, yes, coming from the same boy. This time he took Alice by the hand, pulling her away from the crowd to an exit, 'Shh-ing' her all the way.

Outside, Alice inquired angrily, 'Why did you pull me out of the exhibit? Who are you, anyway?'

'Hello, Alice. My friends in Quantumland told me to watch for you. I am Amit, the Little Biologist,' the boy explained. Alice thought he looked like a younger version of Goswami. *Maybe he's got the same genes!* Alice laughed inside.

'I didn't want you to get into trouble in there.' Amit said.

'But why should I get in trouble?' Alice asked with surprise.

'You were making heretical statements. You are not supposed to question the doctrines. You are supposed to learn them, memorize them, cherish them, and defend them. If your skepticism became known, you would be unwelcome here. You may even be banished from Biologyland.'

'Surely, you are joking. Only by asking skeptical questions does one learn,' Alice said firmly.

'Ideally, yes,' said Amit. 'But Biologyland was beset by too many controversies. They began to get out of hand, so the decision was made to outlaw controversy through indoctrination. You believe in those doctrines, or else you are thrown out of Biologyland.'

'Really? There is no opposition in Biologyland?'

'There is,' said Amit. 'The opposition is called Creationism.'

'I want to know about their doctrines as well,' said Alice. 'Do they have booths too?'

'Booths? No, not here.' Amit was surprised at her question, 'Nobody takes them that seriously.'

'Then how can I meet them?' asked Alice.

'If you insist, I'll take you to one of their rallies,' said Amit with a shrug. 'Come on Alice, let's go!' And off they went.

Presently, they came to a mound of ground where somebody was speaking, and a few scattered people were listening. Or were they heckling? Soon Alice realized it was the latter. Regardless, she moved in closer to the speaker to hear him out.

'There is no evolution,' the speaker was saying. 'God created all creatures, great and small, in one Divine miracle six-thousand years ago in six glorious days.'

'But that's religious dogma,' said Alice. 'How can that be offered as scientific opposition?'

'Exactly,' asserted Amit. 'It can't. But permitting Creationism, clearly religious dogma, to pose as opposition, the ruling hierarchy of Biologyland has managed to obscure the dogmatic nature of its own doctrines. They establish a false dichotomy which is not mutually exclusive.'

'But don't the creationists understand that their outlandish ideas are being used to strengthen the ruling hierarchy?'

'Some of them do. I have heard many people say that the idea of six days of creation is just a metaphor. The main point is that God intervenes in evolution and creates all life, not all at once, but gradually.' Amit spoke with sincere earnestness. 'It may appear in the fossil records to resemble gradual evolution, but underneath, it's all God's creation.'

Alice was quizzical. 'Do I detect a note of sympathy there? Aren't you going to get in trouble yourself if you have hidden sympathy for the opposition?'

'Oh, I am just a little biologist with no professional standing or academic approval, even though I have lots of knowledge. The big fish don't care about my opinions so long I keep them to myself.'

They walked along silently for a while, watching all the people file in and out of the booths of Biologyland.

'You said people who don't swallow biological dogma are thrown out of Biologyland. Is that literally true?' Alice suddenly wanted to know.

'Follow me,' Amit shrugged and led her some distance before speaking.

'Here is the boundary of Biologyland. You can see the big wall,' he said with some fear and resignation.

But Alice saw no tall walls. When she shared her observation with Amit, the poor little fellow seemed even more afraid.

'You don't see the wall because you don't have the eyes of a biologist, not yet anyway. You are a newcomer, not yet indoctrinated into True Biology. When you are properly indoctrinated, your improved senses will make the walls quite visible, believe me,' Amit insisted.

'Yes, like the emperor's new clothes,' thought Alice. But she did not overtly contradict Amit. 'When you are indoctrinated, do you become a citizen and get to vote?' she asked instead.

'No, no,' Amit was quick to answer. 'There is no vote in Biologyland. Only the seasoned experts decide things. But when you are properly indoctrinated, you get to work. You get research grants and government contracts and I want to work.'

'What do you want to do?'

'Many ordinary biologists used to work on the genome project that fully codified the human genome. The genome is the totality of all the genes contained in a single set of chromosomes, you know.' Little Biologist sounded proud of his knowledge.

'But since the Human Genome Project has finished, what's next?'

'We can study the genomes of all the other animals. That should keep us busy for a long, long time. Busy-ness is good for us. An idle brain is the Devil's workshop. And when the Devil gets you, you become a creationist.'

'I'd like to talk to some of those biologists with heretical ideas. May I? Surely, there are some,' insisted Alice.

'But nobody is supposed to speak to them.'

'But there must be a way to know their views; the views for which they became pariahs.'

'There is. But it is too risky. You have to find inconsequential journals, and only crazies called New Agers stock them in their bookstores and post them on the Internet. I suspect those places are under surveillance by the black-ops bio-cops,' said Amit.

'But we are little people. I am Little Alice and you are Little Biologist,' Alice argued. 'You said it yourself. Nobody ever pays attention to little people.'

'I hope you and I are right,' said Amit.

Presently, they came to an alley at the end of which was tucked a bookstore. They went inside. The storeowner moved toward them.

'Hello,' said Amit to the man, 'this is Alice and she's interested in the latest biological heresies.'

'I carry all the New Age biology literature you want,' said the store-owner proudly. 'The Theory of Punctuated Equilibrium, Gaia theory, Morphogenetic Fields, Morphic Resonance, and all the rest.' These exotic terms excited Alice's curiosity.

'I will leave you with him, then,' said Amit sadly. 'I don't dare hear these heresies. They'll only take me off the plantation and confuse me in my life and work. I'm not free to permit this to happen. I need to make a living.'

Alice felt a little saddened, too, as her companion departed. But the lure of forbidden knowledge had her full attention. And the storeowner eagerly responded to Alice's obvious intellectual excitement.

'The current paradigm of biology has reached a crossroads. If it continues in its materialist way, if it continues to assert that life is nothing but a dance of atoms and that our body-mind is a machine, then biology, the science of life, will soon reach a dead end. Biology needs a change of path, or else...'

'Or else it will be pathological for biology, eh?' Alice chortled. The man nodded agreement, somewhat sadly.

'Oh, I'm sorry, but I couldn't resist the pun,' Alice apologized. 'So, tell me some of your New Biology stuff before I read any of the literature. I am only a beginner, I need an introduction.'

'I will do even better,' said the owner. 'Before some of these mavericks became pariahs and went underground, they had a conference with a quantum physicist. Somebody made a video of it that now has become a classic. If you will come into the back room with me, I'll show you the video.'

The phrase 'quantum physics' increased Alice's excitement. Perhaps now she would be able to put together an image of how thoughtful electrons give rise to life. Unhesitatingly, she followed the storeowner to the back room and settled herself before the TV as he turned on the VCR.

On the screen Alice saw a number of people sitting around a large table. She recognized Goswami immediately by his big friendly grin. There was a large motherly woman in a green and brown patchwork dress who called herself 'Mother Gaia'.

'How nice to see a woman sitting with the scientists,' Alice thought. 'I am rather tired of listening just to men.'

The people in the video were discussing the nature of life. Alice blinked and instantly entered the scene on the screen. And she heard about the Morphogenetic Field—an invisible vital field of primary forms that guides life's progress. Goswami explained that it was a little like a habit when the more you did something the stronger it became.

'Life's like that too,' said Goswami, 'as things grow and move they are being guided by these fields of habit, these Morphogentic fields.' And Alice heard about Punctuated Equilibrium—a sort of mysterious quantum jump beyond the bounds of Darwinian gradualism.

And Goswami spoke too, of how consciousness was always working behind the scenes, first bringing matter into existence and then working within life to manifest its higher purposes. And then it was Gaia's turn.

Gaia praised the men for their courage in challenging the current biological theories. 'One of life's finest features,' she said, 'is its urge to transcend barriers, the urge to move beyond elementary particles of matter into atoms, atoms into molecules, beyond molecules into cells. You men are part of that initial stark,' she complimented, 'the same impulse that raised life up out of the primal ooze that moved fishes onto the land and filled the skies with birds, moves you to create theories of how life began, and to speculate about how life takes its form and its direction.'

As Gaia spoke, Alice began to fall into a reverie, as on that sunny day long ago when she and her sisters sat along the banks of the Thames listening to the funny storyteller unfold delicious adventures, poking at pomposity with his wits, and exciting the little ladies to tears of laughter.

She felt herself drifting aimlessly in a dark warm sea. She passively followed the flow of the current letting it take her where it would. As she relaxed she lost all sense of her body and could not tell where she ended and the current began. Nor could she tell how large she was. For all she knew, lazily drifting, she might be as small as an atom or as large as the universe.

Far away she seemed to hear the voice of Mother Gaia speaking, or was it Goswami? What was the voice saying? She could barely hear it:

'And the Spirit of God brooded over the waters,' it said.

'And the One Mind awoke and looked for another,' it said.

'Pull yourself together, Alice,' it said.

And Alice's mind moved in the dark sea and invented atoms, and assembling the atoms, made molecules. Separating herself from the darkness she made dollhouses from her molecules and inhabited every one. Biologists would call them cells, but she called them 'Alice.'

And playing with the little houses she created villages that had a life of their own which she also called Alice. And which biologists would call tissue, would call organ, would call 'organism.'

And Alice stretched herself into a million billion forms, changed her mind and a million forms vanished, to be replaced by a million others. There were forms that swam and forms that flew and forms that lived on light and forms that loved the darkness. And Alice saw that it was good. And she cherished every one of the forms that her mind had created, even the ones that were lost forever. For she remembered everything.

And then all the Alices devoured one another. She did so enjoy eating! Life is so delectable! And they were so much fun to catch. And Alice invented sex for the fun of it, not because she had to, but for the joining and the play. It was almost as fun as eating. And she called that Alice too.

Then Alice invented insects, and elephants, invented writing and fire and photosynthesis. Alice invented seeing and hearing and a billion other senses and enjoyed the world through every one of them. And Alice invented pain and music and movable type. And Alice saw that it was very good.

Alice wept for joy when she invented man, just as she had wept when she invented electrons and the genetic code. Then Alice invented mathematics, granite, steam engines and *The Chandogya Upanishad*. She was always coming up with something new. Her creativity was inexhaustible. Nothing could stop her. And she called it all 'Alice.'

When she had invented quantum theory it reminded her of that dark stream where she had first discovered herself, where she had broken herself into pieces for the sake of the world. Alice wondered if she would ever be able to stop her ceaseless creation of lovely new forms, whether she would ever stop eating. She had enjoyed herself immensely but she was getting so tired and she longed to rest from her play.

Then Alice began to get sleepy and forget her creations and as her mind turned away from them they vanished into the darkness. One by one all the things of the Earth disappeared, and all of the ideas too until there was nothing but Alice alone dreaming beside a river where once the world had been.

And then she saw the Cheshire Cat's reflection in the river, licking its paw.

'Now, Mister Cat of Quantumland, it must have been either me or the Red King or Goswami. He was part of my dream and I was part of his dream, too! Do help to settle it! I'm sure your paw can wait!' But the cat only began on the other paw, and pretended it hadn't heard the question.

Whose dream do *you* think it was?

14 ALICE IN THE BIOLIGHT COMMUNITY

Beverly Rubik

In which Alice joins in the Weak-Photon mediated Dance of Life.

'Oh Quantum Cat, I'm getting all confused again. Why can't I be in some nice and simple place? I'd like it to be how it was when we were sitting beside the river bank.'

The cat looked up at Alice for a moment, 'I think what you want is another good dose of Mother Nature. Just blink once and you'll be there.'

Alice blinked and found herself beside a tall wooden fence with a sign: CAUTION: BIOENHANCIVE RADIATION. Curious, as always, she peeked through a hole in the fence.

'Amazing!' she exclaimed.

But what was it? What had she seen?

On the other side was a garden where the plants and animals seemed to glow softly with a light all their own! Intrigued, she walked along looking for a gate. Stopping periodically to peek through other holes in the fence, she was enchanted by the beauty of the glowing flora and fauna. After walking for several miles she grew tired and discouraged about finding a way in. Struggling to absorb more of the lush radiance by moving from peephole to peephole, she nearly squashed her nose against a large green caterpillar which had just exited one of the peepholes in the fence.

'Aaaaaggghhh!' he screamed.

'Eeeeeek!' Alice recoiled, frightened by his shiny, iridescent green body so close to her face. Since she could not find the gate, she envied this little being, who could move freely through the peephole.

'There is no gate,' he suddenly blurted out.

'What!?! Did you say that?' she shrieked.

'No gate at all. You have to create your own gateway, you do.'

Alice rubbed her eyes in disbelief and weariness.

'Make my own gate, indeed! It's too much for me to explain, but I've come a long way, and I really want to get inside and take a look around before I leave. And besides, I've no energy to make a hole big enough for me in the fence. I'm tired, and I need a good nap to rest after all my walking. But first, please, tell me how I can get in there,' she begged.

'This fence is only the boundary, the edge of your world,' the caterpillar said.

'What nonsense!' Alice replied. 'Humans build fences to keep other humans out—and animals and creatures in,' she reasoned. 'How can this be a boundary of the human world?'

'Most humans, just like cows who follow the herd, don't want to come in here. Too disturbing!' he muttered, inching away from her.

'Wait! Don't go away! What do you mean too disturbing?' Alice pleaded.

The caterpillar slithered up onto a large mushroom nearby, where he reclined on his side, and curled up in the shade of a big oak tree.

An exasperated Alice crouched down on her hands and knees to examine this talking, mind-reading, tiny creature more closely. She looked at him and silently thought...I must be losing my mind! Here I am talking to a worm! I used to dissect worms and look at their tiny beating hearts under the microscope in biology class! Slimy, low, disgusting creatures!... Then she said aloud, 'Who are you, anyway?'

The green worm squirmed as he sensed her unspoken thoughts. Then he suddenly uncoiled and swelled up, enraged. His face flashed yellow like a light bulb and a streak of light ran down his body. His rear end glowed red like a caboose. Menacing, wiggly black horns slithered out at both ends of his body.

Alice retreated in awe and horror at this display.

'I'm no worm,' he retorted. 'I'm the precursor of a psychedelic, pleomorphic butterfly! Why, you're nothing but an anthropocentric, juvenile delinquent!'

'I am not!' Alice retorted, realizing she didn't know the caterpillar's name. 'Excuse me, but we haven't had a proper introduction. I'm Alice.'

'Malice, did you say? It certainly fits!'

'No, Alice! *Alice,* you idiot!' she said, stomping her foot in anger. And he began to move off.

A moment later, she stammered, 'Look, I'm sorry. I've been traveling around so long all by myself, and I've seen such strange things and met such weird creatures, I'm confused. This is not the world I came from. I want to go home, back to normal life.'

'Yes, I know you do. But you can't go back now.' Smug as a bug, the caterpillar grinned a grin so wide, it wrapped all the way around his head.

Alice wondered why the top of his head didn't fall off with such a grin as this! Finally, his message struck her deeply.

'What do you mean—I can't go back?' looking at the fence and thinking perhaps she was in a very strange prison. All of a sudden, her chin began to quiver, and she began to moan. The next moment, she collapsed beside the caterpillar, crying bitter tears.

'Do you mean—I'm stuck here—forever?' she gurgled out, choking on her tears and words.

Feeling compassion for the girl, the caterpillar slithered nearer to her on the mushroom, and said: 'No, no, you're not stuck, and you're not dying, Alice. And nothing is forever! Everything is right now!'

Alice brightened at the thought.

'I'm not? It is?' Alice said, sitting up and paying attention.

'That's right. You're back in Quantumland, Alice, where, as you must know, particles shift around, disappear, and new ones are born each instant. Your mind and body slowly but constantly shift, too, never returning to the same state, but always retaining their form. This change is orchestrated very carefully to preserve the whole, so it is never too abrupt, although it may be uncomfortable at times. There is a new you being born in each instant. Yet, know that you are alive and well and experiencing the vital field which is rapidly changing you in every moment. Once your perception has shifted to embrace Quantum Reality, the world you knew will never be the same comfortable old place.'

Alice pinched herself, still not convinced she was alive. 'But why are there no other people like me here?' she asked.

'Millions of humans think they want to be transformed, reformed, enlightened beings. They'd like to be psychics, mystics, or magicians performing quantum tricks, but they do not understand the consequences. They get stuck, or in trouble, long before they come as far along as you have. That's why we hung up that caution sign on the fence: Bioenhancive Radiation. "Too much life" is another way of saying the same thing. Make no mistake, life is risky and rather than take those risks, many people run away from the source of their vitality. They never come to realize their full human potential. They are content to function safely and mechanically mimic the classical machines they praise.'

'How will I know when I am transformed?' Alice asked.

'When your old questions just fade away and you start asking yourself really new questions; when you start living out of your new view of reality and your new relationships to others. But you will never stop transforming; it is an ongoing process. There are only two things for certain here; one is change, and the other is uncertainty. Or, to think of it another way. Possibility. Freedom.'

'But, what does that mean? How can you be certain about uncertainty?'

'That, my dear Alice, is the nature of Quantum Reality. Now I'd like to show you something. Here, move closer to me,' he coaxed.

The caterpillar moved closer and climbed onto Alice's cheek. She felt his tiny rhythmic feet crawling on her face, and it tickled, but she didn't shoo him away. She sensed that something was going to happen as he perched up on her forehead. When he curled up between her eyebrows, she felt a tingling sensation, warmth radiating, and even more.

'Suddenly I feel—as if—I can read you—like an extension of my own body!' she exclaimed. 'I feel your feelings in the pit of my stomach, knowing you will soon change into a new form, and you are—no, not exactly afraid— but anxious. Very soon now, you will go through a great transformation. I remember this from biology class: caterpillars change into butterflies, and they never return to their old form! But now I am feeling this from inside! I sense your energy and your thoughts and feelings in my head, and also in my heart and stomach. I'm trembling now—with your anticipation!'

'That's right! Very good, Alice! You are learning *bioresonance*. By the way, my name is Luc.'

'Pleased to meet you, Luc.'

'Relax now, Alice, close your eyes, and try this! I will put a flower near your ear. Now, tell me what color it is—and no peeking!'

'What!?'

'See with your ears, taste with your nose, smell with your eyes, and hear with your toes!' Luc sang.

'Luc, you're making fun of me!'

'No, Alice. Just a little ditty to help you remember that your ears are antennae not only for sound, but for many more subtle energies as well. You felt my energy between your eyes, and in your heart and stomach, and so you should be ready now to read with your ears!'

'Read with my ears?'

'Come on Alice, repeat after me: see with your ears, taste with your nose, smell with your eyes, and hear with your toes!'

'Are you crazy?'

'Vibrations are emitted from the life all around you, but you humans tune out or ignore 99 percent of the *bioinformation*! Instead, you pollute the environment with radar, radio, television, and other junk waves that have no life value, only commercial value. Here, we tune in to *geo-cosmo-biorhythms*, and we are one with the whole community of life. Now close your eyes, Alice.'

He plucked a Tiger Lily with his sharp choppers and tickled her ear with it.

'Ouch, that hurt!' the flower said. Then in a moment, the flower sang in multiple tones, 'What color am I, hmmmmmmm?'

Alice's ears began to burn. 'Red? I'm just guessing though. Are you really red?'

'Well, I'm orange, a fiery reddish orange, though. Pretty good, Alice!' the Tiger Lily answered.

The caterpillar picked another flower and held it near her ear. 'Can you feel it?' he asked.

'I feel many vibrations and hear many voices singing. What does that mean?'

'This flower is white,' said Luc. 'White, of course, has all the colors of the rainbow. Many hear it as a chorus singing, and sometimes as the hiss of a white water stream rushing by. Flowers also emit *biophotons* of various colors.'

'What's a biophoton?'

'The extremely weak but coherent, aligned, wavicles or particle-waves emitted by all life forms. Biophotons contain vital information about a life form's state of health and its needs. In the garden over there beyond the fence, everyone is glowing and reading everyone else's glow in an endless conversation of biophoton emission and absorption. We all absorb and beam out *electromagnetic bioinformation*! It's the quantum glue of the biosphere!'

'Electromagnetic what?'

'Electromagnetic bioinformation. It's the subtle information vital to life that is coded and carried by biophotons—particles of light radiated and absorbed by all life forms. The bioinformation is heard by some as a very complex conversation. They hear many voices speaking on top of one another. But we do try to be coherent: emitting and absorbing in phase with each other, contributing to the unity and well-being of the whole biosphere, _Ecologically Enhancing the Energy_, or EEEing others by serving the quantum whole of life. When we experience the bliss and harmony of the coherent field of life all around us, then our own life is enhanced, too.'

Luc continued: 'Back in your old classical world humans behave as isolated things, competing, struggling, fighting Nature, fighting themselves, hardly aware of the true Quantum Nature of Nature, or of their own full potential in macroscopic, large scale, quantum coherence. Aside from a few individuals that they label as "psychic" or "cuckoo," humans are completely insensitive to subtle life energies and bioinformation. But you, my dear Alice, are displaying some openness, awe, and wonder about life beyond the conventional human notions of reality. You're no longer just a material girl. You're becoming a *thoroughly Quantum Girl*. You are becoming sensitive to, learning to read and resonate with, the weakest biofields. Now, if you still want to visit the garden, you can go in...'

'But how can I get into there, Luc?'

'Visualize yourself in the garden as if the fence were no barrier at all.'

'What!? How can visualizing get me in there?'

'In the quantum state, what appear as barriers, are not really barriers at all. Quantum beings—as shimmering waveforms—can tunnel right through them. Imagination and visualization are the keys to unlocking choices in material reality. Come on, let's go! I'll see you through it.'

Alice stood up, facing the fence, but a few cloud-like doubts continued to loom in her mind. Nonetheless, she decided to follow Luc's instructions and give it a go. She closed her eyes and started to visualize traveling through the fence into the garden. In her mind's eye, she saw herself become a liquid flowing through the holes in the fence, like water through a sieve, merging into a stream again on the other side.

'No, that's not quite right,' said Luc, perceiving her mental image through bioresonance. 'Too classical a picture. And you must banish all doubts! See yourself already on the other side, without any barrier, and in no time at all!'

Alice tried visualizing the fence disappearing, and disappear it did. For a brief moment, she felt herself de-localize into no-body-at-all, and suddenly, there she was again, a collapsed quantum wave function Wiffed! into the luminous garden, full of a variety of radiant beings: trees, flowers, insects, birds, rabbits, squirrels, and many other creatures.

Alice was astonished that she had accomplished this new form of transportation.

'Whew, for a moment I felt I had no body at all, but here I am again!' She laughed with relief. 'Anyway, Luc, I'm in here at last!'

Luc watched Alice as she began to prance around, to dance out of pure joy. All the creatures came to watch her, attracted to her girlish energy and exuberance. The energy field of the community enveloped her, bathing her in pulsations of love and joy. As the field grew stronger, she became dizzy, intoxicated from its intensity. Entering a coherent excitation, she danced wildly, propelled by the pulsating biofield. Then, layer by layer, all the creatures joined in. They swayed and moved in step with each other, making a complex dynamical, self-organizing pattern that evolved all by itself, like a symphony being conducted without a maestro. It was a perfect picture of quantum coherence at the macroscopic level!

Much to her amazement, Alice found that now, while dancing within the biolight community, she could do perfect pirouettes, effortlessly every time! It was as though the coherent field of dancing creatures allowed her to move in perfect dynamical form with hardly any work at all.

She took a great leap, over and above the community of creatures who had gathered to watch, spreading her legs like the wings of a great eagle at an angle of 180 degrees in the air, performing a glorious grand jeté. All the creatures gazed up at her with so much love and light—with their biophotons pumping up her energy field so high—that she was able to soar like a bird, suspended way above the crowd for a considerable time, as if she were dancing in outer space without any gravity. Finally, after several minutes, she floated down and landed gracefully in the biofield that was by now, humming, glowing, and as thick as molasses.

'This is it! I can absorb the cosmic life energy from all around me. I'm able to move by this inner light, transforming it into the dance, and then giving the gift of this energy back to everyone,' Alice shrieked in delight, 'this is *the* dance!'

'Sacré bleu!' cried Luc, as aqua-colored-caterpillar tears of joy streamed down his face; his red taillights pulsated in unison, and antennae at both ends vibrated in ecstasy. 'Now you've got it, Alice!'

Bioluminescent sparks flew out from her fingertips and toes as she twirled and leaped. The entire community broke into dance with her in radiant vitality and *joie de vivre*. The scene looked like a coherent display of spinning fireworks. On this evening, on this magical afternoon, all was super radiant!

As darkness fell upon them, the dance relaxed to a ground state of slow, sustained gyrations and oscillations, and then to a quantum jiggle that was barely perceptible in the growing stillness of the night. A chorus of frogs began to croak Brahms' 'Lullaby' in unison. Alice lay down to sleep on a mattress of soft, dense moss. All the creatures drifted off into the night to rest.

In the morning, Luc presented Alice with the gift of a toy called 'pet particles in a box.' Some of the time, some of the particles were visible, and at other times new ones appeared. The combinations of particles seemed unpredictable, endlessly creative, and they never repeated a pattern. The particles themselves were charming, colorful, spinning, and quantum-correlated, dancing around inside the box, and sometimes outside. As she watched their strange antics, she learned that she could play with them through her own conscious intent.

'Although they are only particles,' she noticed, 'they behave a bit like that, my dear cat. Sometimes when I just think about him, he comes over to me, apparently reading my mind. And so do these particles! So I guess there must be Mind in matter at all levels, even in the most elementary particles!'

'Indeed!' Luc replied, pleased with her conclusion. 'The visible effects of the invisible manifest themselves in everything, even particles. Particles may be elementary in the sense that they are the primary physical stuff of life, and everything else in the universe, but they are far from simple. And an atom is an amazingly complicated, finely balanced assemblage of forces and particles woven together in exquisite detail. Small is not necessarily simple.'

'So I see, Luc' cried a delighted Alice, coaxing her toy particles to move in little figure eights.

All of a sudden, Alice noticed a disturbance on a nearby hillside.

'By the way, Luc, what is all that commotion over there?'

'A fluctuation broke out. Would you like to help quell it?' Luc asked.

'What do you mean?'

'The fluctuation started with a little boy lamb, who experienced a quantum quirk in the night, and was very disturbed,' Luc reported. 'Now the whole herd of sheep is upset. We could join the EEE squad Alice, to practice our ecological energy enhancement to soothe him, and smooth it out to the benefit of all. Do you want to give your new powers a try?'

'Okay, let's go.' She bent down to pick up Luc. He slithered into her hand, glowing all the while, and off they went.

The little lamb was shaking and bleating incessantly, and all the sheep were quite agitated. The whole community of other creatures had gathered to help the lamb, and they were busy making final preparations for a sacred, quantum ritual.

Ignoring them, Alice ran over to the little lamb, and cried, 'Here, let me hold you!' She lay down with her arms around the poor creature. She cradled her hands around his head and neck and held him softly, in a loving, warm embrace, trying to comfort him, stroking his thick white fleece.

'Look, he's better already!' she said.

The little lamb stopped bleating, and looked up at her with his doleful, sweet eyes.

'Why that's strange! What is happening?' asked Alice. 'My hands feel like they are burning with fire, deep inside my bones!'

Observing this, Luc said: 'Your hands were made to do many great things, and one of them is to touch others with love and compassion and send them healing energy. Within the whole kingdom of animals, human hands are extraordinary, not only for their dexterity, but because they are remarkable antennae sensitive to the entire spectrum of life's subtle energies. They are also powerful radiators of electromagnetic bioinformation. You were the perfect conduit for sending healing energy to the little lamb. Why, just look at them!'

Alice's hands were pink and swollen, hot and heavy with bioenergy carrying love and intention to heal. In her heart, she had felt the pain of this poor lamb, who had been spooked by a nightmare, and now she felt peace that he was resting comfortably among his kin on the hillside.

The community of creatures made a large circle as part of their ritual. By making contact with each other and with Alice, they formed a bioenergy circuit. Alice reflected on how enriched, expanded, relaxed, and full of peace and love she felt in this circle of life and light and love.

'I feel like my entire being, in all its dimensions, is part of you, and you are part of me. I feel the power of the clan, the quantum community. I am so, so much more than just Alice!'

'And so you are, our dear, Alice!' said all the creatures in one harmonic voice. 'We are coherent. Look, we are glowing brighter and brighter! We are light, we are one, we are super radiant!'

Luc suddenly jumped up into her hand.

'Alice,' he said, 'this morning I shed my last skin. My work as a caterpillar is finished. I am ready to transform. Please, help me through it.'

'Yes, of course, Luc. But what should I do?'

'Can you please let me do it right here in the palm of your hand, so full of light and love. Oooooohhh, I feel it coming on already! Here goes! I am oozing! I am...'

While Alice and the whole community watched her outstretched hand, he melted down into a puddle without visible form. The Luc she knew was seemingly gone, and Alice cried out in dismay. The only thing left of him was his glow and a powerful psychic sense that Luc was still there, somewhere, but not in his usual biological form.

'Luc, what is happening? This is not how caterpillars normally change!'

But Luc was not there to hear or speak about it. He was a green puddle of liquid flesh, cradled in the palm of her hand. The puddle started to oscillate in various modes. The whole puddle moved in phase. First there was one mode—the whole puddle throbbing in unison—then two—the puddle breathing like two fists alternately opening and closing—then four--pulsing like the chambers of the heart—then eight, then sixteen, and then more modes than anybody could count, and they oscillated faster and faster. The green puddle of Luc had become a moving vortex of fluid so energetic, that it lifted itself up and began to spin around as if on an invisible axis—like a gyroscope without any point of support! Then it looked like a small tornado of light just above the palm of Alice's hand. Everyone watched expectantly, wondering just what would happen next.

The vortex spun for several minutes, as it gradually changed color, from iridescent green, through a color wheel of different shades, and finally became a flaming purplish-pink. Then the spinning vortex began to slow down. Amazingly, parts of the radiant beauty that were the new Luc started to become visible to Alice.

'Luc, are you okay? How do you feel? You are—why, why, you're a beautiful butterfly!' gasped Alice.

Luc said nothing. He was dizzy, dazzled, and be-frazzled. He tried fanning his new pink and purple wings, and at the same time he struggled to extend his long spindly legs. Wet and wobbly, with wings now spread, he made an effort to stand up straight in the palm of Alice's hand, but toppled over.

Alice saw that he was uncomfortable. Sensing his uncertainty about whether he liked his new form, she cradled him gently between her hands and sent him love, hope, and prayers with her biofield.

A few moments later, Alice felt something change, and she opened her hands to take a look. There was that enormous grin of old Luc, wrapped around his needle-thin, butterfly head!

'Haaaa-haaaaa! Heeeee-heeee-heeee-heeee-heeee!' he laughed with glee.

'Luc, what's so funny? We're so concerned about you! Are you okay?' asked a perplexed Alice.

'I am reborn! After what I have been through, I appreciate as never before the wonder of the life force within us all!'

'Please, tell us more! What happened?' cried the quantum community in unison.

'I surrendered to the living matrix,' answered Luc.

'What?' asked Alice.

'I have a new mission,' said the transformed Luc, proudly fluttering his beautiful wings. 'Flowers everywhere will embrace me as I suck their delicious nectar, and at the same time, I will pollinate them to become fruits. Animals will eat the fruits, spread their seeds, propagating more plants. I am delighted to nurture the web of all my relations.'

'This is great news!' shouted the quantum community.

'And yet, someday, I know I will become food for the birds!' He winked, and went on, 'That is the ultimate service, one's own sacrifice in harmony with the cycles of Nature! But the beauty and grace of life is never lost! And so be it! I now take my adult place as a butterfly in the Great Chain of Being!'

An uproarious call for celebration erupted in the quantum community: 'Come on! Let's celebrate!'

Then Luc joined the others in a spontaneous parade honoring his transformation.

'I am grateful to you, Alice,' Luc cooed, held high by the parading crowd, 'for helping me through my transformation with your wonderful love-filled hands. Come on Alice! Let's dance!'

Luc flew in graceful loops and figure eights around Alice's body, topping off his performance by weaving a sparkling gold luminescent crown into the beams of blue biolight pulsing around her head.

'Wow! Look at him go! What derring-do!' the crowd cheered.

Alice and all the creatures were overwhelmed by the new high frequencies of joy and love that Luc beamed to them. As one great blissful being, they absorbed and beamed new variations of love and joy back into the universe. EEEing with its powerful message, that ultra-weak radiation, was part of the cosmic life force that is everywhere.

With tears in her eyes, Alice beamed at the brand-new butterfly: 'And I am grateful to you, too, Luc, for helping me get past that fence. Now let's do the quantum jitterbug!'

'*EEE-HAAA!*'

15 ALICE UNDERGROUND

Nick Herbert

In which Alice visits the Deep Reality laboratory.

'That was a very nice adventure indeed,' Alice said to the now familiar feline smile next to the fence. As she did so, the Quantum Cat phased into view. 'But to tell you the truth, I'm getting a little homesick. Do you think it would be very nice to see Oxford again.'

The cat stretched himself and said, 'Oh, the place is no problem, just not too sure about the time, but that won't matter because you're at home in many places and many times—and not!'

In but a blink Alice found herself on St. Giles Street following a crowd into one of the colleges. Inside was a large banner which read: 'Oxford University Presents: A Panorama of Modern Science.' From the crowd's style of dress Alice concluded that she was in one of Oxfords most prestigious, modern science buildings. The audience was filled with famous astronomers, cosmologists, mathematicians, and particle physicists from all over the world.

The first speaker was a short dark-skinned man with black hair wearing a blue suit and red tie. 'Good afternoon, ladies and gentlemen. Welcome to the Panorama of Modern Science. My name is Raja Bucke and I am from India, a mathematical physicist with many degrees and discoveries to my credit. This afternoon, it will be my pleasure to explain to you some of our discoveries and to answer any questions you may have about the workings of modern science.'

The audience applauded, respectfully.

'Thank you, ladies and gentlemen. I am reminded that science begins whenever we search for explanations for what is happening around us. But not all explanations are equal: some explanations are simply wrong. For thousands of years people tried vainly to explain the world and its affairs as the acts of supernatural beings. Everything that happened in the world, they guessed, was the result of some intention: human intention for human acts, divine intention for cosmic acts. And in between? For intermediate activities between divine and human, people's imaginations invented a variety of beings—angels, devils, elemental spirits, hobgoblins and fairies whose less-than-divine wills made the ordinary world go. But something important happened 300 years ago that banished every one of these imaginary beings from the real world and sent them back into the story books where they rightly belong.'

Alice nodded her agreement.

'The Age of Superstition ended when Isaac Newton discovered, while at Cambridge's Trinity College, that the world is governed in every detail not by spiritual intentions but by impersonal mathematical laws. Newton discovered some of these laws, and his successors—Maxwell, Einstein, Heisenberg, Schrödinger, and Dirac discovered many more. Now we scientists possess a mathematical structure orbiting around quantum theory that correctly describes every natural phenomenon that we can measure, to an accuracy of a dozen decimal places. Quantum theory has been tested in every way and in every place and in every way we know how, and its predictions have never been wrong. Today we no longer appeal to the gods to learn how the universe operates. We merely write down and solve an appropriate mathematical equation. Forgive me for boasting, but for the first time in human history, science seems to be able to answer any question we can ask.'

Raja from behind the podium did a little jig, singing in the style of Gilbert & Sullivan:

We've got a quantum formula
The best that's known to man
And all the world's phenomena
Calculate we can.
Yes all the world's phenomena
Calculate we can.

Alice's eyebrows rose a little at the man's thoroughly un-academic behavior, and was even more surprised by the audience's warm applause.

Raja returned to the podium and continued: 'But science does not rely on mathematics alone. Science stands on two feet: theory and experiment. And here to tell you about science's experimental side is my esteemed colleague Buck Roget.'

Roget stepped to the podium, embraced Raja, to warm applause, and began to speak: 'Greetings to you all. I am Buck Roget from the University of Paris. I am distantly related to the famous thesaurus-maker and I have an important physics laboratory located within view of the Eiffel Tower. Raja has told you about the significance of theory in science. I represent science's other foot—the experimentalist.'

'One of the hallmarks of the Age of Superstition was that so much of what they thought they knew was based on hearsay, on what they had read in so-called "sacred books," or on the word of some authority. In science our motto is: "Question Authority." Trust only what you can see and hear with your own eyes: trust in experiment. If you can measure it in a laboratory, it's true. If you can't measure it, it's probably nonsense.'

'Raja will tell you that there *is* an infinite number of possible mathematical theories that might explain the world. My job as an experimentalist is to make measurements of the world that will disprove most of these theories. The ones—or one—that remain unrefuted we will accept as our current explanation of the world.'

'I will have to tell you from the start there is no experiment that can ever verify a theory because there may be many other theories that could explain the same facts. But it takes only one experiment to refute a theory.'

'And as Raja says: only one theory that we know of passes all our tests. Since its formulation in 1925, quantum theory has withstood the most determined efforts of several generations of scientists to refute it. It is the most successful theory of natural phenomena that we have ever known.'

'Besides explaining present facts, this theory predicts unexpected phenomena which lead to new technologies which enable us to do previously unthinkable experiments which deepen our knowledge and confidence in the theory. Some of the new technologies built on quantum ideas are atomic energy, semiconductors, superconducting magnets, and lasers.'

Alice thought, *Why he's making all the quantum madness sound so normal!*

'Our new quantum technologies allow us to see further out into space and deeper into the structure of matter than ever before. Soon we will be able to measure everything that can be measured. Soon nothing will be hidden from the eyes of the quantum experimentalist.'

Alice suddenly remembered Rosie's analogy of scientists acting like little boys looking up Nature's skirts. She giggled.

On stage now, Roget was dancing arm-in-arm with Bucke, singing:

We've built a high technology
that we can really trust
and all the world's phenomena
measure it, we must
Yes, all the world's phenomena
measure it we must.

They're both being so silly! Alice thought. And once again, the audience expressed its warm appreciation.

They returned to the podium and Raja announced that, before the break, they would answer a few questions from the audience.

Alice listened to some of the questions and finally became bold enough to ask one of her own. 'Sirs,' she said, 'you say that science stands on two feet—experiment and theory—but what about *reality*? What does science tell us about the way the world *really* is?'

After an embarrassing silence, Raja spoke. 'A very good question, Miss. We scientists do not usually think such deep thoughts. As a scientist I would say that I know nothing about reality: I am only acquainted with theories about phenomena. But if I were pressed to answer your question, I would say that reality is mathematics, that the mathematical patterns are real. What would you say about reality, Doctor Roget?' He turned to his colleague.

'If you can measure it, it's real. If you can't, it's not. Later in this talk I will show you some of the wonderfully real things we can measure. But for now, thank you all for your questions. We will begin again in thirty minutes. Tea will be served on the First Floor.'

Alice wandered out into the hallway, avoiding the crowd, disappointed with Buck and Raja's answers to her question about reality. Three phrases continued to echo inside her head: 'Reality is Mathematics. Reality is Measurement. Tea will be served on the First Floor.'

She walked into an open elevator at the end of the hall and pushed the bottom button. She noticed too late that it was labeled MINUS ONE.

That's odd, she thought, as she looked closely at the elevator buttons: MAIN FLOOR, PLUS ONE and MINUS ONE. Why don't they just call it the basement, instead of Minus One? Alice wondered impatiently. I bet that tea is being served on PLUS ONE.

Before Alice could change her mind, the elevator landed. When the door opened, Alice could see a wooden sign on the wall directly opposite the elevator: 'DEEP REALITY LAB: Getting to the Bottom of Things.'

She stepped into a long corridor going to the left and the right with many small doors on either side. 'Maybe I will skip the tea,' said Alice as the elevator door closed behind her with a thud of finality.

But before she could make up her mind in which direction to go, she noticed to her left a large bald-headed man with a tremendous handlebar moustache wearing a white lab coat and pushing a tea cart toward one of the doors. 'Excuse me, Sir,' she called, 'Is this the Reality Lab?'

'Why, yes it is, Miss,' the man said politely. 'But what are you doing down here? We get so few visitors these days—everyone upstairs is so interested in mathematics and measurement. Nobody really cares about reality itself. You must have gotten into the old freight elevator by mistake. Would you like to join us for tea? My name is Doctor Walrus and you are...?'

'Walrus?' Alice was startled because he looked like one! 'Very pleased to meet you, Doctor Walrus. My name is Alice. Are you really looking for reality down here in the basement of the science building?'

'Yes, we have been looking for reality for quite a long time. But this is not the basement of the science building. Officially we don't exist at all. Reality Lab is not on any map of Oxford. It is staffed by eccentrics, mavericks and misfits, and privately funded by the super-secret Charles Lutwidge Dodgson Foundation...'

Alice started at the mention of Dodgson. 'Oh, wouldn't you know it! Rev. Dodgson funding puzzles and games for eccentrics, mavericks and misfits, with one called "Walrus,"' she giggled under her breath.

'One of the conditions of our presence at this college is that we keep out of sight and not embarrass the university,' Doctor Walrus lamented. 'Come inside and meet my colleague Professor Carpenter.'

Alice entered an office lined with books and framed awards. A tall, thin, bespectacled man named Professor Carpenter greeted Alice warmly. He was dressed in blue overalls, and his many pockets were filled with tape measures, rules, compasses, chalked string and a variety of other instruments of the builder's trade.

They all sat down and were momentarily silent as Walrus set out the cups and poured each of them a cup of Earl Grey tea. Before sipping his tea, Walrus reached into his pocket, extracted a box of snuff, shoved a pinch of it up his nostrils, then, turning his head to one side, emitted an explosive, very wet sneeze.

'Oh, that feels good,' he exclaimed. 'This Copenhagen snuff really cleans out my brain. I couldn't think clearly without it. Can you believe our good luck, Carpenter? This bright young lady is actually interested in reality? What a find!'

'And what did you learn upstairs, Alice, about the nature of reality?' asked Carpenter.

'Well,' Alice began, 'Raja said that reality is mere mathematics and Buck said that reality is only what you can measure.' She leaned forward in anticipation and asked, 'What do you two think about that?'

Walrus stared thoughtfully into his cup of tea and finally said: 'Well I believe that they are both half right.'

'And both half wrong,' exclaimed Carpenter.

'Yes, the success of quantum mathematics in predicting the results of every experiment we are clever enough to devise certainly shows that the mathematics is revealing some great truth about Nature,' said Doctor Walrus. 'But I personally believe that this mathematics does not represent reality itself, but only describes the structure of our human transaction with reality.'

Alice recognized this immediately as Niels Bohr's model of quantum reality. 'So you would agree with Niels Bohr's view that "there is no deep, hidden reality" beyond the physical that we can ever know? That only physical phenomena are *real*?'

'My, my. You've been learning quite a bit on your adventures. Yes,' the Walrus concurred, 'I also believe, like Bohr, that, when taken together, the quantum mathematics and the quantum facts point to a reality that lies forever beyond human description. Quantum reality is a realm we were never meant to inhabit. We simply do not have the physical and mental equipment to experience such a reality—even in our imaginations. As far as we humans are concerned, there is no quantum reality. It is fruitless to speculate about its nature. Scientists must learn humility in the face of Nature's mystery. For me, reality research is a search for new facets of this mystery that I can appreciate, while at the same time knowing that I will never be able to grasp reality completely.'

'Doctor Walrus, haven't I seen your picture in the newspapers? And isn't that a Nobel Prize medal in the cabinet among all the snuff boxes?' asked Alice.

'Yes, it is, my dear,' Walrus said.

'Why, you're no maverick scientist. You're one of the world's top theoretical physicists. And yet you claim there's no reality down there? Just mystery?'

Embarrassed by the attention, Walrus sat in silence. It was Carpenter who spoke. 'You're right, Alice, Walrus is one of the few top scientists interested in the reality question,' said Carpenter. 'Most of the others are working on theory and experiment just like Raja and Buck and the rest of those guys upstairs in the Panorama of Modern Science. Down here we believe that "Reality is the real business of science." Albert Einstein said that, by the way, and Einstein was a true seeker after reality. Theory and experiment are necessary tools he thought, but what he really wanted to know—behind all that high technology and complicated arithmetic—is what this world is really like.'

'Then what's your own opinion, Professor Carpenter, on the reality question?' asked Alice, truly interested.

'Well, I'd have to agree with Walrus that deep down the world is unthinkably strange, so strange that humans will never be able to imagine it. But wherever and whenever humans choose to interact with that deep and utterly mysterious realm, we must always get a clear and common-sense result—a result we can describe in ordinary language.'

'Oh, so you would agree with Dr. Bohr, and Dr. Heisenberg, who believed that sub-atomic phenomena cannot be expressed in ordinary language because atoms aren't things? Are you one of the Copenhagen boys I've been warned about?' Alice wanted to know exactly where Carpenter stood.

'Yes, I agree with Bohr and Heisenberg,' the Carpenter answered.

'But, you're saying that every quantum experiment that we can perform on an atom can be described in ordinary language but the atom cannot be so described. How can that be, Professor Carpenter?' asked Alice.

'The results of each experiment that we are able to perform on quantum reality are all perfectly ordinary—when taken one at a time. But the results of all experiments taken together point to an utterly mysterious whole. It's like the story of the blind man and the elephant. Each one of our quantum measurements makes perfect sense, but when we try to imagine a reality that could give rise to all our results taken together, we are baffled. Yes, our mathematics predicts these results, but there seems to be no quantum elephant behind the scenes that would give us a conceptual picture about how these particular results could possibly come into being.'

'So you agree with Buck Roget that only measurements are real?' Alice asked with disappointment.

'Yes,' said Carpenter, 'the world becomes real only as the result of some measurement.'

'But,' interjected Walrus, 'the mathematics, although not reality itself—as Raja Bucke seems to think—points however imperfectly to a deeper reality—a reality that is however, in my opinion, beyond the ability of humans to comprehend or to even imagine.'

'How can you be so confident,' asked Alice, 'that scientists of the future with expanded imaginations will not be able to picture a deeper reality with metaphors currently unavailable to twentieth-century minds?'

Walrus took another pinch of snuff from his box and, after another explosive, wet sneeze, continued: 'Well, it is of course impossible to predict what people in future times will be able to do. I speak only of our present way of thinking. Given all I have learned about physics, the philosophy I have studied, and all the logic I have imbibed at the universities, I see not the slightest glimmer of hope in being able to describe quantum reality in terms comprehensible to human minds.'

'I remember long walks in the park with Carpenter where we continually asked one another: Can Nature really be as queer as she seemed to be in our early experiments with atoms? I can still remember all those exciting results we began to uncover after Heisenberg's wonderful discovery of quantum mechanics in 1925. And after nearly a century of quantum experiments and Herr Schrödinger's invention of a marvelous mathematical structure which predicts these experiments exactly, we can now say with confidence: This world is a very queer place indeed! Whether self-made or created by God, this universe is certainly made according to a decidedly non-human, construction plan. Whatever we may learn in the future, we now know that for a fact. Quantum queerness is here to stay.'

'Queerer than a three-legged duck,' added Carpenter. 'Just ask Dodgson next door. He's been working on something he calls "Quantum Logic"—trying to speak the unspeakable by changing the very rules of reason! Poor fellow's very nearly mad. Goes around reciting nonsense verse. Just go right in. He'll appreciate your company. We mostly avoid him. I'm afraid that even among reality researchers Quantum Logic has a bit of a bad name, goes a little too far over the edge.'

'Well, maybe he can help me then with my search for what's real. I'm pleased to have met you both,' said Alice politely. 'Now, which way to Quantum Logic?'

'It's to your left as you go out the door,' responded Walrus.

'And it's to your right if you back out,' continued Carpenter. 'Very nice to have met you, Alice. Come back again.'

'Kachoo!' exploded Walrus from behind a friendly cloud of tobacco-scented mist.

Back in the corridor, Alice knocked tentatively on a door marked 'Daniel Dodgson, D.D.' until she heard a polite voice say: 'Enter, please.'

Inside she discovered a barefoot young man dressed in an orange fluorescent jumpsuit with his long brown hair in a ponytail. He was seated on the floor of a room completely empty of furniture, whose walls were covered with pictures of impossible figures and optical illusions. On the floor around Dodgson lay odd-shaped wooden blocks and plastic objects with the same paradoxical character as the pictures on the walls.

'Hello,' said the man in orange from the floor. 'Its name is Dodgson, D.D. Pray what in this world are thee?' he asked in a singsong falsetto.

'Alice is my name, Sir. May I sit down and join you?'

You may if you can
My pretty young man
But now my knees see
That Alice is she.

'Mr. Dodgson, please! I thought you were investigating new ways of thinking about reality!'

'OK, Alice, I'll talk in prose. I only pretend to be crazy to frighten off the merely curious. But I know that you are a serious student of science.'

'How do you know that? Can you read my mind?' asked Alice rather sharply.

'No, the walls here are very thin. I was eavesdropping on your conversation next door.'

'What kind of research are you doing, Dr. Dodgson? I understand it's called "Quantum Logic." Is that the purpose of all these funny shaped toys on the floor?'

'Just call me Danny, Alice. And "toys" is exactly what they are. Quantum toys. I agree with Walrus that human imagination cannot grasp this world's deep reality. So I'm trying to unlearn my human sense of the world that I learned uncritically as a child and teach myself a new quantum way of thinking. My idea is that the quantum dilemma might be solved by making a radical change in our very laws of thought. Perhaps the world really consists of atoms whose positions are always definite—hence no measurement problem—but we can only properly talk about these atomic positions using new grammatical rules for combining the words "and," "or," and "not." That's what these toys are for. That's what these pictures are all about. I'm trying to reprogram my thinking so that quantum paradox becomes completely ordinary.'

'And have you succeeded?' Alice inquired.

'Yes, to some extent. I am beginning to think like an atom. My brain is becoming quantum-logical. Imagine a quantum zoo that contains only four kinds of animals: cows and horses each of which is either all white or all black. Now imagine a big field with two gates. At one gate I let in only cows. At the other gate I let out only white animals. Of course, I find that half of these white animals are horses. In the quantum zoo at any one time you can specify either the color or the species of an animal but never both,' Danny explained.

'But what does that have to do with atomic particles,' she asked.

'It's the same with atoms: an atom's x-spin and y-spin obey exactly the same logic as the quantum horses and cows. Unlike classical physics—where a particle could possess all its attributes at the same time—quantum attributes always come in what's called "incompatible pairs," like position and momentum, energy and time, or x-spin and y-spin. Once you choose to know the value of one attribute, its partner attribute becomes completely uncertain. Incompatible attributes are just a "way of life" for an atom or a photon, and that's the way I'm trying to learn to think.'

Alice remembered hearing about Heisenberg's Uncertainty Principle and was intrigued by the possibility of Quantum Logic.

'So does your new way of thinking give you a new understanding of reality?'

'Yes it does,' Danny answered with some excitement. 'Now I can grasp quantum reality in a way no human being has ever done. But there are several problems with this approach to quantum reality research, Alice.'

'Tell me, Danny, tell me about your problems.' Alice sat down and cocked her head to one side, as Danny continued: 'First of all, Alice, it is impossible for me to share my new insights with my colleagues without reprogramming their brains, too! It took me almost ten years to think like an atom—I still can't think like a molecule—and I was highly motivated. I can't expect a skeptic or even a friendly peer to invest that much time on what, to them, is such a most dubious enterprise. So, my insights into deep reality belong to me alone. One of the main features of scientific knowledge is that it can be shared. I fear that what I am doing might not really be science. But science or not, I insist that my path be regarded as a legitimate form of knowledge.'

'Well, I can empathize with you, Danny. It seems you've carved out a most difficult arena for exploration,' Alice said, trying to build up his self-confidence

'A second and perhaps more serious problem,' he went on, 'is that when I am thinking like an atom, the ordinary world becomes paradoxical. Looked at through quantum-logic goggles, this familiar, phenomenal world right where we are, makes no sense at all! Yet surely the universe is a logical unity. There cannot be one logic for atoms and another logic for apples.'

'It looks like you can't help me, Danny,' Alice sighed, 'and I can't help you. I don't have ten years to spend learning a logic that's useless in every-day life. But could you give this busy girl at least one unforgettable kinder-garten lesson in your new logic before she goes?'

'I'd be delighted to, Alice! I've never had a more enthusiastic pupil, even though, you are my first student,' Danny admitted, laughing. 'Now, just take off those clumsy shoes and follow me into the rumpus room.'

Alice quickly removed her hiking boots and followed Danny through a door labeled: 'QUANTUM LOGIC LAB: NO SMOKING'. Inside it was completely dark except for the light coming through the doorway. Alice was startled as the door slammed shut behind her.

'Are you ready, Alice?' Danny called out from somewhere in the darkness.

'Yes, I think so,' she answered a bit timidly.

She heard a tiny click in the corner and suddenly the room was filled with a brilliant light, then darkness, then light. Danny had turned on a powerful strobe light, and during its brief flashes, Alice could see that the 'Logic Lab' re-sembled a large dance studio with a hardwood floor, cushions and prop boxes along the edges plus one wall that was mirrored from floor to ceiling.

Alice had always liked strobe lights, and now she pranced across the floor watching her mirror image fractured into stop-motion snapshots—like frames cut out of a movie. Danny jumped up from the corner and joined her on the floor, and together they played in the strobe light, holding hands and whirling, glancing at their chopped-up images in the mirror—a barefoot boy in an orange jumpsuit dancing with a striped-stockinged girl in a blue-and-white pinafore.

'This is my first quantum lesson, Alice,' said Danny, a little out of breath as they continued to move about in the flickering light. 'The world pops into existence when it's observed, dissolves into "something else" when not observed. The images we see in the mirror obey ordinary logic, but in the darkness, Quantum Logic rules.'

'Can you play some quantum music for me, Danny?' gasped Alice, breathless from the thrill of the dance.

Danny stopped for a moment, and turned on his favorite quantum-log-ical disco beat. 'I like to imagine that in the dark, in between flashes, I'm like Schrödinger's Cat, a mysterious being made of pure possibility,' shouted

Danny as they danced. And not only possibility, but WAVES of possibility. I'm experimenting with different kinds of music—music that, for me at least, evokes the fertile nature of quantum darkness.'

Alice was quite excited by thinking of herself as waves of possibility in the darkness, then as an actual 'Alice' in the light.

'Oh, Danny, this is Wonderful. I could dance all night!' Alice exclaimed in pure delight.

'Not in kindergarten, Alice. Let's play ball instead,' Danny said as he reached into a prop box, extracted a large silver beach ball and tossed it to her. The ball's strobe-fractured trajectory made it difficult to manipulate, but soon they were bouncing the ball between them with ease in the flickering light!

'Now bounce it into the mirror, Alice,' called Danny. And when she did, TWO balls bounced back! Danny caught one and Alice the other.

'There's one example of quantum logic. In the quantum darkness ONE Plus ONE can sometimes add up to FOUR. What the physicists call CONSTRUCTIVE INTERFERENCE of possibility waves. Here I've just simulated that with a magic trick.'

'FOUR?' asked Alice. 'Oh you mean the two balls in our hands and the two in the mirror.'

'Yes, you've got it, Alice! Now let's try some more mirror bounces,' he suggested.

Alice bounced her strobe-lit ball a few more times against the mirror, but it just returned in the ordinary way. Then her ball hit the mirror and suddenly vanished!

'ONE plus ONE equals ZERO. DESTRUCTIVE INTERFERENCE of possibility waves,' announced Danny, tossing Alice his own silver beach ball. 'Zero's the other extreme: when two equal possibilities meet in the dark, because they are waves they can add up to any number between zero and four. In the quantum world one plus one is rarely two.'

'Playing strobe-ball like this is teaching me quantum mathematics firsthand,' Alice gushed.

'But that's probably enough for your first lesson, Alice. The strobe is beginning to give me a headache.'

Danny put on some more restful music, dowsed the strobe, and turned on the main lights. He leaned back on a large cushion by the wall. Alice walked over in her stocking feet, put down the ball, and joined him.

'I like what you're doing, Danny,' Alice said breathlessly. 'But suppose you succeed in programming your mind to think quantumly? Won't everyone in the real world think you're insane?' she asked intently, curling up closer to him.

'Well at least I'm trying to do something about the quantum puzzles, not just talk about them. And I don't think I'm insane, not yet anyway. Do you think I'm insane, Alice?'

'Why no, Danny, I don't. I just think you're naturally curious,' Alice reassured him.

'But there's one guy down here who's truly out of his mind,' said Danny. 'He's trying to exist in a million different stories at the same time.'

'You mean Walrus and Carpenter?'

'No, Alice. This fellow lives down on Minus Two. He's actually not even a human being, but an alien life form stranded here some time in the last century. Walrus told you about the Dodgson Foundation and the need for secrecy. Charles Lutwidge Dodgson, by the way, was my great-great-great uncle. Well knowing about "Kat," as we call him/her—its real name is unpronounceable—is really hush-hush. It'd really embarrass the university if the press found out Oxford's been harboring an alien for more than a hundred years! And besides, Kat values his privacy, too. He wants nothing more than to be left alone with his pipe and slippers to work on his manuscript about the nature of deep reality. He's already written hundreds of pages, all in Pleidean of course! So Kat and I have a similar problem: How to communicate our insights about the quantum world to other human beings.'

Alice's curiosity was piqued: 'Where's Minus Two? Can we get there from here?'

'We would both find it quite uncomfortable, Alice. Minus Two is an underground railway tank car filled with high pressure Argon and Nitrous Oxide gases. We can communicate with Kat only through closed circuit TV. In my great uncle's day they used Morse code, then telephone. Because of my interest in alien logics, I'm one of the few folks here with a TV hookup to Minus Two. Would you like to look in on Kat with me?' he asked, knowing she would.

Without waiting for a response, Dodgson threw a concealed switch and the mirrored wall became a giant TV screen, showing mostly swirling clouds of vapor.

'Here, let me turn up the contrast,' said Dodgson. Out of the fog emerged an enormous blue caterpillar, sitting on a red-and-white mushroom, smoking a hookah.

'That picture!' exclaimed Alice, sitting up suddenly. 'It's straight out of *Alice's Adventures in Wonderland*!'

'Yes, you're right. Charles Lutwidge Dodgson sketched—and Sir John Tenniel drew—Kat from *real* life, not their imaginations. Kat's ship crash-landed near Oxford in 1864. That mushroom-shaped seat is Kat's "slipper." It puts him in instant mental contact with all others of his species. Caterpillar says he is actually part of a giant group mind that's been exploring this part of the galaxy for millions of years.'

'So what's he smoking?'

'The substance in the hookah? That's a secret between him and Walrus. It's apparently derived from a plant that grows only on Earth. It's the main reason Kat is still on this planet. He calls it "Wisdom Weed" and claims that many galactic races less civilized than Pleideans would invade and enslave us to gain access to such a divine commodity.'

'Oh my, perish the thought,' said Alice, a bit frightened.

'Oh, by the way, he can see us too. This is a two-way connection,' said Danny.

Caterpillar's head slowly turned as if awakening from a trance. 'Hello, Danny,' droned the deep synthetic-sounding voice.

'And who are you?' Kat intoned, twisting its body to look directly at Alice with its million-year-old eyes.

'I, I, I,' she stammered, 'I want to know more about reality. My name is Alice,' she finished lamely.

'Reality is my passion also. My name is (*indecipherable squawks*). But you can call me Kat. You've heard all about Hugh Everett and the idea that there are many, many universes that are created each time you make a quantum measurement.'

'Oh yes, my cat told me all about it. In fact I've been to some of those universes already.'

'By the way, it may interest you to know that I consider quantum theory to be one of the most beautiful cultural accomplishments of your race. It is most like what we Pleideans call (*indecipherable squawk*), but executed with a particularly poignant human touch. I scan your modern physics books with a pleasure that is almost sexual. It makes me Squawk Squawk as does nothing else on Earth with the exception of Wisdom Weed. I love your quaint physics. It is a small but delightful window into one corner of deep reality.'

'On behalf of everyone in Quantumland, let me say thank you for your appreciation, Kat,' Dodgson said.

Cat continued with his assessment. 'Quantum theory, as you know, represents the world as billions of vibratory possibilities only one of which becomes actualized during measurement—according to the conventional view. Everett wondered, as a graduate student at Princeton, just what is so special about that one possibility that allowed it to actualize? Why not imagine that all possibilities are created equal, and that each has an equivalent 'right to life'? Why not imagine that all possibilities—not just one—are actualized simultaneously, each in its own separate universe?'

'Well, why not?' asked Alice. 'It only seems like the democratic thing to do.'

'Everett's Many-Worlds Model of quantum theory is the most audacious story I have yet encountered since my brief contact with your species—we have been visiting you for less than 700,000 years,' informed Kat. 'We Pleideans are very interested in stories. Our space ships are in some sense powered more by stories than by physics.'

'But stories are just in the mind,' claimed Alice. 'How can a story drive a space ship?'

'There is a particularly haunting phrase from Everett's work that moves me greatly. May I share it with you? Replying to his critics who objected to his gratuitous multiplication of universes, Everett maintained that the universe really doesn't split at all; what splits is the observer. Here is that ominous phrase: 'It is not so much the system which is affected by an observation,' said Everett, 'as the observer who becomes correlated to the system."

'So did he mean a new way of "looking" was needed?'

'Yes,' said Kat. 'Now, may I please tell you a story inspired by those words? I will try not to frighten you.'

'Kat's stories are very persuasive,' said Danny turning to Alice. 'But I'm always wiser afterwards. Let's listen, shall we?' He slipped a comforting arm around her shoulders.

'OK, Kat, tell us your tale,' said Alice, who was smiling up at Danny.

'Settle yourself comfortably on your cushions, close your eyes, and attend to my voice. And now we will begin,' said the ancient being in the Argon tank.

Alice and Danny sat down.

'Once upon a time there were two Curiosities named Alice and Danny. I call them Curiosities because they were both intensely curious about the mystery that surrounds us, the mystery that was invisible to most of their friends.'

'Danny Curiosity was very excited because he had just received a gift from his friend Kat—a being even more deeply curious than Danny. It was a little black plastic box that Kat called an "Everyscope"—which was something like a kaleidoscope but much more—shall we say, "engaging?" The insides of the box were very simple: it contained nothing but a very weak light source and a half-silvered mirror. Kat assured Danny that this toy would teach him more about quantum reality in a few minutes than he had learned in his whole lifetime.'

Alice listened to Kat with rapt attention.

'Danny was anxious to share his gift with his new friend Alice Curiosity. Danny guessed that Alice would also enjoy the toy, and he wanted to surprise her with the magical present from his alien friend. He wanted Alice to be with him when he tried it out for the first time.'

As she listened to Kat's voice, Alice was surprised to find herself drawn into the story in a most vivid way, as if she were experiencing an intense dream. She was barely aware of sitting on her cushion; most of her was in another place altogether, fascinated by the curious black plastic box that Danny was holding in his hand.

'The light source is very weak,' Danny was saying. 'It emits only one photon per second. That single photon strikes the mirror and half of its possibility wave is reflected back, and the other half goes through. The reflected wave comes out this hole marked "X" and the transmitted wave comes out this hole marked "O." And that's all there is to Kat's so-called "Everyscope."'

'How does it work? Did it come with instructions?' she asked him.

'Yes. It says on the back in gold letters; now watch both holes closely. Which one lights up?'

'Is that all? That doesn't sound very complicated,' Alice frowned.

'Oh yes, there's more on this side. It's that quote from Hugh Everett that Kat is so fond of: *It is not so much the system that is affected by an observation, as the observer who becomes correlated to the system.* And on the other side it says: *Be all that you can be.*'

Danny turned to Alice. 'Is that one of Buddha's sayings, or is it from Krishnamurti?'

'I really don't know, Danny. But, why don't you go first? I'll just watch,' she responded cautiously.

Danny held the 'Everyscope' in his hand and looked at both holes. Suddenly he saw the X-hole light up. And an identical Danny right next to him saw the O-hole light up. Each of the two identical Dannys was holding an Everyscope and in the next second each of them split in two making four in all. After five seconds there were 32 Dannys in the room.

'Wow!!!!' the 32 identical Alices said, watching this performance. 'This cat box is turning us both into Quantum Hussies! What does it feel like, Danny, when you split in two like that?'

'It feels beamish, Alice. Would you like to try it too?'

Then the 32 Dannys simultaneously handed their Kat boxes to the 32 Alices.

'Let me see,' said the 32 Alices. 'Yes, the X-hole lights up.'

'No,' said 32 extra Alices, 'the O-hole lights up.'

'Oh this is so much fun!' said all the Alices at once. 'It's absolutely frabjous.' By now there were 4096 Alices and 4096 identical Dannys in the room.

Alice and Danny Curiosity continued to play with the magic box for hours exploring the ever-new experiences it brought them and their numerous identical offspring. And they all learned ever so much on that magical day, the day of Kat's present, the great Everyscope day, and everybody lived with everybuddy else and they all lived happily ever after. THE END!'

'THE END!' proclaimed the Kat in the Tank a second time, waking Alice and Danny from their trance.

'My, that was a marvelous story. It seemed so real,' said Alice. 'You are a wonderful story-teller, Kat.'

'Thank you, Alice. That is the greatest compliment a member of my species can receive.'

'But back to reality, Great Teller of Tales,' said Alice. 'Do you believe that the Everett Model is real?'

'It's very believable, Alice, a very plausible yarn. Very insightful for a race so young.'

'But if the "Many-Worlds Interpretation" is a true picture of reality,' 'argued Alice struggling to keep her composure in the aftermath of Kat's particularly compelling story, 'then what is the significance of this one world, the one that we seem to be meeting in now?'

'It is of no significance whatsoever. All worlds are created equal.'

'But do you really feel that, Kat? Or do you feel like I do, that *this* existence is something special, because it is the one that happens to be happening to *me*?'

'Yes I feel like you, young Earthling female, that this reality is special. But my intellect tells me that my feeling may be mistaken. Feelings often are, you know? I was stranded on this planet by a misinterpreted feeling.'

'Really?' interjected Danny. 'I thought that your ship crash-landed.'

'I let you believe that. But I was marooned on this planet by forces that were more psychological than physical.'

'What really happened, Kat, in the summer of 1864?' probed Alice.

'Our crew was routinely observing your planet, as we have done for many millennia, when I inadvertently violated a minor point of sexual etiquette vis-a-vis the group commander and I was sentenced here. Our customs are too complicated to explain here: we have 23 or more different sexes, the number depending on your definition of "sex" and "desire." On many of the Everett worlds this unfortunate accident did not happen, and in those I am still traveling with my crew harvesting new stories and engaging in the various emotional and physical entanglements available to members of a group mind almost a billion years old.'

'Do you miss your crew, Kat?' Alice asked sympathetically.

'Not much. My stay here is but a day in a very long life. My comrades have already forgiven my transgression. I laugh about their prank. They will be arriving momentarily—on my time scale not yours—to rescue me. I have enjoyed sharing this time with such a young and hopeful species. For me this visit has been very restful compared with space travel—I have had so much time to think. So nice to meet you, Alice.'

'Nice to meet you, too, Kat,' said Alice.

'But do seriously consider the Everett story,' Kat continued. 'I believe it to contain a portion of truth. Now, if you will excuse me, I must return to my manuscript. Give my fondest regards to Walrus and the rest of my friends on Minus One, and convey my highest appreciation to the trustees of the Dodgson Foundation who have treated me so well. I will not forget their generosity.'

The screen went black.

'Did that really happen?' said Alice to Danny, a bit stunned by her encounter with the philosophical caterpillar from the Pleiades.

'Yes, Kat is real, all right. But you don't have to buy his philosophy, Alice, just because he's a billion years old. In some ways science on the Pleiades is less advanced then our own. They abandoned the study of matter long ago in favor of research on something they call 'sfonk'—the nature of which Kat has never been able to explain to me even when I'm thinking like an atom.'

'Thank you, Danny, but I think I've had enough of your friend Kat to last me ten lifetimes. I think I'd like to go home now and find my real cat. She told me what to do, just close my eyes and blink'.

PART II:
QUANTUM SPECULATIONS & JOURNEYS TO THE OTHER SIDE

16 A PHOTON'S EYE VIEW

Peter Russell

In which, in Conversation with a Photon, Alice Learns what it Feels Like to be a Particle of Light.

And so Alice closed her eyes and was about to blink again when...

'Good morning, Alice,' a voice said. Or at least it seemed like a voice.

Alice rubbed her eyes and looked around. She wasn't home at all but in a garden. And where had that voice come from for there was no one in sight. Had the voice been in the dream, she wondered? She'd caught herself enough times in that trap, thinking she's been awake, only to discover she was dreaming. And it always annoyed her.

'Good morning, Alice.'

There it was again. But where was it coming from? Alice had become used to voices that came from strange and unexpected places, or were disconnected from the people or things that were speaking, but not voices that came from absolutely nowhere.

'Good morning,' replied Alice cautiously, but politely, not wanting to upset whoever, or whatever, this might be. 'Who are you? Or more to the point, where are you?'

'I'm your actual quantum,' the voice continued. 'You've been hearing a lot about quantum physics and all the strange conclusions that it leads to in your world, so I thought it was time you heard from me, and got a picture of how the world looks from a quantum's point of view. As to where I am, I am everywhere and nowhere. Always and no-when.'

Alice knew better than to let her mind get bothered by paradox. Just about everything she had heard so far was paradoxical in some way or other, and trying to understand them was bound to lead to even greater confusion.

'Let me introduce myself,' it continued, 'and all the other zillions of quanta in the Universe, for in many ways we are all exactly the same. As you know, each of us is the smallest possible packet of energy in the universe. Any transfer of energy, whether it be from one electron to another in an atom, or from the sun to your skin, involves a whole number of us quanta. It may be 1, 2, 5, 117, or 19387463728 of us, but never half a quantum or three-and-a-quarter quanta. That would be like you having a conversation with half a person, or three-and-a-quarter people.'

Alice wondered whether she could imagine having a conversation with three-and-a-quarter people. Three-and-a-quarter bodies, perhaps—she'd met stranger situations than that. But three-and-a-quarter people, she was not so sure. But before she had a chance to try imagining a fraction of a person, the voice from nowhere was back.

'And in ways some might think egocentric, I am a very special quantum, qualitatively different from any other quanta. In your world you also call us photons—the smallest unit of light. By the way, when I speak of light, I am talking not just of the visible light that you see with your eyes; I mean the whole spectrum of electromagnetic radiation of which visible light is just one tiny range of frequencies. At higher frequencies are ultraviolet light, X-rays and, beyond them, gamma rays. At lower frequencies you find heat waves, and at the lowest frequencies of all, radio waves. All of them are just different frequencies of light. And they are all composed of photons, each one a single quantum.'

'Then why did you say you were all the same?' Alice asked. 'Light has many different colors; heat I can feel on my skin; and I've been told to keep well clear of gamma rays. They seem very different to me.'

'That is because the energies we carry vary enormously. The higher the frequency, the higher the energy. A gamma-ray photon, for example, packs billions of times more energy than a radio-wave photon. This is why gamma rays, X-rays, and even ultraviolet rays to some extent, can be so dangerous to you. When these photons hit your body, the energy released can blow apart the molecules in a cell. When a heat radiation photon is absorbed by your skin, the energy released is much, much less, and all it does is warm you up a little. But although our energies vary enormously, there is something about us that is always the same. We all, each and every one of us, possess exactly the same amount of action.'

What, Alice was about to say, is action? But before she had even finished thinking...

'What is action?' the quantum repeated. 'I thought you might ask that. You're familiar with the terms such as mass, velocity, momentum and energy, I presume?'

'Yes', thought Alice. She remembered learning about them at school.

'And you learned how they relate to each other? An object's momentum, for example, is its mass multiplied by its velocity. And work is energy multiplied by distance. Action is just another one of these qualities, but it is not one you normally hear about at school. The amount of "action" in any action is defined as the object's momentum multiplied by the distance it travels. Or it can also be expressed as the object's energy multiplied by the time it is traveling. Imagine someone throwing an apple, for instance.'

The White Rabbit ran across the grass, throwing apples into the air.

'If he threw twice as fast would you say there was more or less action?'

'More, of course.'

'Twice as much?'

'Probably.'

'And if the apples were much heavier, like croquet balls, would there were be more or less action in his action?'

'More.'

'And if he ran around for twice as long, how much action, do you think there'd be?'

'Twice as much, I would think.'

'So it's not really that strange is it?'

'No,' replied Alice, wondering why she had never thought about action as a measurable thing before. And why she hadn't heard about it at school. Maybe it hadn't been important.

'Oh, it's very important,' the voice from nowhere responded. 'Your mathematicians have discovered that whatever happens in the universe, happens in such a way that the total amount of action is always the lowest possible. It's what they call "The Principle of Least Action." And your scientists use it all the time to predict how things will happen. Those apples the White Rabbit was throwing traced out a curve in the air, yes? Well that curve happens to be the particular one that involves the least amount of action. Any other curve you could imagine would require more action.'

A sort of cosmic efficiency principle, thought Alice.

'Yes. And it applies to everything. Even light. When you see your reflection in the looking-glass, the light comes back to you at that precise angle which involves the least amount of action.'

'I can begin to see why action is so important.'

'Yes, it's absolutely fundamental. And, as I was saying, every single quantum in the universe, every photon whatever its frequency and energy, is an identical unit of action. The amount is exceedingly small, after all we're very, very, very tiny. In your units of measurement, each of us is about 0.0000000000000000000000000663 erg.seconds. And before you even think of asking what an erg is, it is a unit of energy. But a very small one. To lift one foot off the ground, a croquet ball weighing one pound, takes about a thirteen-and-a-half million ergs. And an erg.second is a unit of action. If you took a second to lift up the croquet ball, your action would have involved about thirteen-and-a-half million erg.seconds. Now each quantum is a tiny, tiny, tiny fraction of an erg.second-point zero zero zero zero zero zero zero zero...'

'Stop, please. I get the picture. You are a very, very, very, tiny unit of action.'

'Yes, the smallest possible action in the universe. It's called Planck's constant, after German physicist Max Planck who first discovered us, almost fifty years after you were created, my dear. Each one of us, each and every one of us, possesses exactly this same amount of action.'

Alice thought about this for a while.

'Light is action,' she mused. 'I'd never thought of it like that before. But I suppose it sort of makes sense. After all, light never stops moving. It can travel right across the universe, and at great speed. And I remember being taught that nothing ever travels faster than light—and then learning from my quantum pals that many "non-things" like entanglements do connect faster. Since light never rests, never slows, action does seems an appropriate name for its essence.'

'Not so fast,' the quantum interrupted. 'That may be how you see light, but we see ourselves very differently. As far as we are concerned, we don't ever experience ourselves traveling anywhere. We never move at all.'

'Now that's ridiculous,' cried Alice. 'Paradoxes, I've got used to in this quantum world of yours, but how can you say you never travel anywhere when you so clearly do? If you never go anywhere, how come light gets to us from the Sun, and how come light has speed?'

'Hold your horses, my dear, and I'll try to explain. But to do so I need to take you on a little excursion into the theories of another of your great scientists, Albert Einstein.

Like many other scientists of his time, Einstein was puzzled by the fact that light always seemed to travel at the same speed, no matter how fast you might be moving. At first this seemed nonsense. If you were to walk along at three mph, and the White Rabbit ran by at seven mph, simple arithmetic tells you he would be going four mph faster than you. If you speeded up and ran along at seven mph you would be able to keep up with him, and there would be no difference in speed. But light didn't seem to behave like this at all. Experiments seemed to show that however fast you went you could never catch up with light; it always passes by at 186,000 miles per second. Even if you were to travel at 185,990 miles per second, light would still whiz by 186,000 miles an hour faster.'

'Faster! Faster!' This time it was the White Queen's voice; and Alice knew where it was coming from. Images of chessboards, talking flowers and railway guards flitted through her mind, and she remembered what it was like to never get anywhere however fast you ran. 'Was the White Queen a friend of yours?'

'No, but maybe young Albert as a boy had read about your adventures with her. After a lot of thought he decided to accept as fact that you could never catch up with light, however fast you went. That is just the way the universe works, however nonsensical it might seem. This led him to his famous "Special Theory of Relativity," and to some conclusions that at first seemed even greater nonsense. His equations predicted that the faster something went the more slowly its clocks would run. The precise relationship between speed and time is not a straightforward one, and I won't bother you with the detailed mathematics, but the result is that if you were to travel past someone at 87 percent the speed of light, they would observe your clocks to be running at half the speed of theirs. This slowing applies not just to clocks, but to all physical processes, all chemical processes, and all biological processes. Your whole world would run at half the rate of theirs.'

'Sounds more like the looking-glass world than my world.'

'Well, it turns out that your world really is a bit like the looking-glass world. Scientists have flown clocks around the world on jet planes and found that they do indeed run slow—by a factor of about one in a trillion—not enough to worry anyone, but enough to prove that Einstein's theory is correct.'

'And it's not just time that shrinks. Space is also changed. Lengths measured in the direction of motion become shorter, and in exactly the same proportion as time slows. So if you were to travel a measured mile at 87 percent the speed of light you'd measure the distance to be only half-a-mile.'

'Is that why it's called "Relativity?"' wondered Alice.

'Yes, space and time turned out not to be as fixed as people had thought. How much space and how much time you observe is relative to your speed.'

'But although space and time had fallen from their absolute status Einstein discovered that not everything about time and space was relative. People moving at different speeds may disagree on the amounts of space and time separating two events, but all observers will agree—no matter how fast they may be moving—on the total amount of space and time between the two events.'

Alice imagined it must be a bit like cutting a string in two. Cutting it in different places would give pieces of differing lengths, but the total length of string would always be the same.

'Exactly. Or rather, not exactly. Space and time don't add up by simple arithmetic. In fact, you get the total by doing a subtraction.'

'Doing addition by subtraction!' objected Alice. 'Sounds like the sort of arithmetic the White Queen would like.'

'Actually it's not quite as simple as that. The mathematical formula for combining space and time is a bit complicated. It's something like "the square root of space squared minus time square."'

'I think I'll skip that. I'm confused enough as it is. But what's all this got to do with light?'

'Well, the equations of Einstein's Theory of Relativity predict that at the speed of light, length will shrink right down to nothing, and time will slow to a complete standstill.'

'You mean space and time just disappear? That is bizarre!'

'Yes, and it's quite troublesome to your physicists because their equations of motion get littered with zeros and infinities, and it's very hard for them to make much use of them. So they usually ignore this extreme case, consoling themselves with the thought that, because nothing can ever actually travel at the speed of light, they don't have to worry about these bizarre limit effects.'

'Why do you say things can't travel at the speed of light?' asked Alice, sensing a possible contradiction.

'Ah, that's because not only space and time change with speed, but so also does mass. But whereas space and time decrease with speed, mass does the opposite. The faster you go, the heavier you become. If you reached the speed of light, your mass would become infinite.'

Alice tried to imagine having an infinite mass. Being very, very heavy she could just about handle. But infinitely heavy? She couldn't even imagine infinity, let alone an infinite amount of anything.

'Don't worry. You'll never get that far. To move an object of infinite mass would take an infinite amount of energy. A lot of energy might get you close to the speed of light, but there simply is not enough energy in the whole universe to accelerate you all the way up to light speed. That's why it's impossible for anything to ever travel at the speed of light.'

'But some things do travel at the speed of light,' interjected Alice, pleased to think she might have caught the photon out. 'You, for example, travel at the speed of light.'

'Of course. To say that light couldn't travel at the speed of light would be pretty ridiculous, wouldn't it? But light is not really a "thing" as you think of things. Photons have no mass at all. Each of us weighs absolutely nothing—no matter how fast we go. Even at the speed of light, we still weigh absolutely nothing. So we aren't subject to the same cosmic speed limit as you are.'

'And so you always travel at the speed of light.' Alice proudly concluded.

'On the contrary. We never travel at any speed.'

'What?' Now Alice was completely befuddled.

'No, that's just how it appears to you in your world. On our side of the quantum looking-glass, things look very different. I said that at the speed of light distance and time shrink right down to zero. Well that means, from our point of view, that we never experience ourselves traveling any distance whatsoever. And, we take no time to do it. Strange as it may seem to you, we never experience ourselves going anywhere at all.'

'Makes the White Queen seem positively rational,' Alice mused.

'The White Queen was still living in the world of things, the world of space, time and matter. We quanta live in a very different world. We are not things. We have no mass, we never travel any distance, and we experience no time. And because we travel no distance in no time, the notion of speed is meaningless for us. In our frame of reference—and what frame of reference could more appropriate for light than our own—we have no need of speed.'

'But I thought Einstein said that the speed of light was the same for all observers. So how can you say that you have no speed?'

'What you think of as the speed of light is from our perspective something very different. You remember me saying that all observers always agree on the total amount of space/time separating two events, even though they disagree on how much actual space and how much actual time they observe?'

'Yes.'

'Well, when you calculate the total amount of space and time between the two ends of a light beam the result is always exactly zero. This is because the total is arrived at by that complicated formula that involves "space-squared minus time-squared." For any photon, anywhere in the universe, the amount of space it appears to travel is exactly balanced by the amount of time it appears to take, and when you subtract the two, they cancel each other out, leaving a total of zero. This is something even we photons agree upon. Except that we don't experience a beginning and an end. We, remember, observe ourselves traveling zero distance in zero time. Subtract zero from zero and what do you get?'

'Zero, of course,' Alice calculated.

'Exactly. However you look at it, the combined amount of space/time for light is always zero.'

The photon continued: 'In your world you observe a separation between the beginning and end of a light beam. This zero amount of space/time has manifested as a certain amount of space and a corresponding amount of time. Since the total must remain zero, the amount of space that appears is exactly balanced by the amount of time that appears.'

'???????'

'What you observe as the speed of light can be thought of as the ratio of manifestation of time and space. For every 186,000 miles of space, there appears one second of time. It is this ratio that is fixed. This is why the so-called "speed" of light in your world is always the same.'

Alice didn't quite know what to think. She sat back and tried to imagine what it would be like to be light. She tried imagining space and time disappearing, but it didn't work. However hard she tried, space and time would not go away. 'Maybe that's just the way the human mind works,' she thought.

Then she tried the opposite, trying to imagine nothing being stretched out into space and time. But that was just as difficult. She was just about to give up trying to understand any of this when suddenly the thought came to her that if light doesn't experience itself traveling anywhere, then what's all this stuff about light being both a wave and a particle? 'It can't be either!' she exclaimed.

'You're catching on fast,' observed the quantum. 'A wave that traveled nowhere would be ridiculous, wouldn't it? So would a particle that existed for no time at all. No, waves and particles are concepts, ideas, thoughts, your scientists use to try and understand us in their world. They are both "thing" words; but we photons, like other quanta aren't things at all. Trying to make us seem like things is one reason your scientists find us so puzzling and paradoxical. They are seeking to understand us from their world, the world of time, space and matter. And that's a big part of their problems. Remember, we photons have no mass; we don't inhabit space and time. We don't belong to the world of things. We belong to the world of light. If they could step into the world of light they would realize that there is no paradox at all.'

Alice lay back, and closed her eyes. There was the White Queen again, with that incessant grin. Or was it the Quantum Cat's grin? And why was the White Queen, or Kat, or whoever it was, wiggling like that? Oh, it must be the Queen, because it's wearing a crown! thought Alice. Her edges grew fuzzy, her colors blurred and her crown began to turn into golden light. Before she knew it, the White Queen had completely dissolved. And so had everything else. There was nothing in her mind but light. No thing. Nowhere that was anywhere else. No time that was not now.

Is this the same light? she wondered. Is the light I see in my mind the same light as the light I see in the world? Is everything light? Is the light of consciousness the light of the world?

'What else? But I think we should leave that for another time,' she heard the voice from Nowhere and Nowhen say. But at least she now knew where the voice was coming from. It was coming from Everywhere—from Everywhere around her, and from Everywhere within her, in know time.

17 CAUGHT IN THE WEB OF TIME

Fred Alan Wolf

In which, from a Time-Weaving Spider,
Alice is introduced to Quantum Speculations
about the Human Soul.

Alice was immersed in Deep Uncertainty. Where was she? How did she get here? Sometimes it appeared to her that she was floating in a white cloud. But when she looked quickly, it seemed that the cloud was not white, but a glistening red or blue. And then, just as she recognized what color it was, the cloud went black and she heard music coming at her from all sides. 'Perhaps the music is only in my head', she thought. Then when she tried to figure out what the music was, she would hear talking, not *words* she could understand, but just the *sound* of voices.

Alice looked around her mutable environment. Suddenly, she felt a definite shiver.

'Curious,' she said, 'I'm shivering and yet I'm not cold, not a single bit.'

Then the shiver occurred again and she could have sworn she saw *herself* out of the corner of her eye off to her extreme right! Then another shiver, and she glimpsed *herself* again! This time to her left. And then with the next quiver she saw herself above *herself*!

'Oh, dear, I'm getting quite beside myself,' she cried out. When she realized what she said, she laughed.

Alice waited for the next quiver and when it came, she quickly looked up and saw herself above herself again, as if she was looking in an overhead mirror. But as soon as she attempted to focus her eyes on herself, her duplicate vanished! Things are becoming *curiouser* and *curiouser*, she thought to herself, especially since her very own thoughts seemed to be not only in her head—but in front of her, beside her, on top of her, to her left, to her right, below her, and in back of her! Oh, my! How could this all be? she wondered to herselves. Then suddenly she understood the source of the shimmering colors.

'*AAAGGGGGHHHH! EEEEEKKKK!*' she cried out loud. She was standing on a strand of a gigantic vibrating spider's web! The strange color changes were apparently due to white light reflecting from the shimmering web. Then when no light reflected from it, it seemed to vanish.

Just then the web shook again and coming right towards her was the largest spider she could ever have imagined, quaking with laughter! In fact when it saw Alice, it laughed so hard, it set the web shaking all over again.

'Well, hello there,' the laughing temporal weaving spider giggled, 'Such a surprise!'

The spider put one of its many arms over its mouth. It was trying hard to keep itself from laughing.

'I haven't caught anything in this web for an Eternity. In fact I haven't caught anything in this web since it was constructed. In fact, I think it was constructed. In fact, I think I did it, but it all depends on how old I am and I can't tell that either. IN FACT, IN FACT, IN FACT, IN FACT!'

The spider was growing as big as its words. When it spoke them, the words sprang out from its mouth, caught in the web, and instantly frozen, with a crackling, shrinking sound. And when the web shook, like broken icicles the words just tinkled from the web, falling away into the darkness. Eventually, the spider calmed down and the web stopped shaking.

'This is certainly the weirdest "dream"—if this is a dream—I have ever had!' Alice decided. 'And you are the weirdest spider I have ever seen. That is, if you really are a spider. Are you?' she wondered.

'I am, I am, I am, I am! Oh dear I'm getting all caught up on my words again. I've just got to stop doing this!' insisted the laughing temporal weaving spider.

Again its words shot out of its mouth and hung themselves on the lines of the web. The spider continued: 'I guess I should tell you what this is all about, but please don't get me too excited because then my words keep getting bigger and sticking to the web, and then I get even more excited and start laughing and then, and then, and then.'

It was happening again. But the spider knew it, so it immediately calmed itself down. 'So if it's all right with you, I'll just whisper.'

Alice nodded her head in agreement as the spider told his story.

'I am the weaver of time, I think!? I began this web a long time ago, just back there,' the laughing temporal weaving spider said as it turned its head off to one side, 'and I finished this web a long time from now, just up there,' as it turned its head up in the direction the light seemed to be coming from. 'This is a very special web as you can see. Because it exists only in time but not in space!'

Alice was getting confused. 'How could that make any sense?' she asked.

'These filaments which make up my web are called timelines and I weave them,' the spider explained. 'I must be very careful however, because the web I set up becomes the Skeleton-Scaffolding of Existence itself. This is the Word-World of Time Without Space. It is the Word-World of If-Ideas, Maybe-Thoughts, and Possibility-Dreams, before they become realities and actions in your familiar world. From these words, ideas, thoughts, and dreams, spring all the things we see about us. Space also springs from these strands and all of the things in existence come from them. Since this is a spaceless space-of-time, it is as big as your head, as small as a dot, and as large as the universe right now; as large as the universe was, and as large as it will be, when it is as big as it will get which is 300,000 times bigger than it is right now, or so the *idea*-theory of General Relativity tells us.'

'Whew! That was a mouthful! Do you mean we are inside of my head, right now?' Alice asked.

'Yes, and we are inside of mine, and we are inside of his head too!' the laughing temporal weaving spider said holding his sides with three of his arms.

'His head?' Alice wondered. 'Whose head is "his?"' Just then she heard what seemed to be the voices she had heard earlier. She could swear she recognized one of the voices.

Why it's Professor Flow! But what is he doing here? Oh dear, for that matter what am I doing here and where is 'here' anyway? she thought. The spider is insisting that 'they' were inside of her head and at the same time 'they' were inside of the universe, and at the same time, 'they' were smaller than a dot!

How could they all fit inside? Alice asked herself. Had they all taken a bite of the shrink-me drink? Or was it a drink of the shrink-me mushroom? It seemed so silly to bite things you drink and drink things you eat, but in this world, words themselves seemed to be changing their meanings. To calm her befuddlement, Alice listened carefully to the professor's words.

'Most people think of the soul in terms of its reality,' Professor Flow began lecturing. But who was he speaking to? He was speaking to an invisible audience! He also was quite animated and intent in his delivery.

'They ask: Is the soul material or an illusion? This natural question unfortunately introduces a gap separating modern science and spiritual thinking and leads to the split situation we presently find ourselves in. We are led to see material things as real and spiritual things as beyond matter. To find the right trail to a new soul vision, we need to look at time in an entirely new way.'

Alice wondered again just whom the professor was talking to. She listened hard and suddenly she heard her own voice answering! It was her voice, but then again, how could it be? She was listening to it as if it was coming from somewhere else, and she was not making up the words! Her voice appeared to be coming from a very, long distance away, and it seemed as if the speaker—which was after all herself—was speaking down a long corridor.

Paying closer attention to Professor Flow's words, Alice thought to herself. What is he talking about? The soul? Why should he be talking about that? He is a scientist not a clergyman. But then again perhaps all scientists are a little clerical in their makeup. But why should he talk about that *now* and why here?

Professor Flow began to answer as if he had heard her question: 'The soul is coming back into fashion. And today we are faced with a new sense of meaning for the soul. We speak of soul-crisis and soul-loss. In this world of compromise, we often think that we must give up our souls to gain a comfortable survival. We once more are a flock looking for its shepherd, a people seeking their souls.'

Alice then heard the blah, blah, of sheep's voices. The sheep went on bleating plaintively.

'Blah, blah, blah. Blah, blah, black me, have I any wool?' the sheep complained, 'All they do is blab, blab, blab. Why don't they blab something "blah-lievable" for a change?'

Just then the sheep disappeared. When Alice looked up, she saw the smiling spider looking down at her, grinning like the Cheshire Cat that vanished leaving only its smile behind.

'I gather you don't like the professor's words,' said the smiling temporal weaving spider playing with three of his legs.

'Oh, no,' Alice denied. 'I love his words.'

'But then why did you bring in the sheep?'

'But, I didn't!!' Alice said, emphatically. But then she thought about it. 'Perhaps I really don't like his words and I only like his magic. Could it be that everything I think here in this strange world just manifests? Oh, dear, I must be very careful not only about what I say, but even what I think!'

'Yes, my dear, you must be very careful which words you choose to use when you think, for here certain words tend to lose their meaning and they cause *immeasurable* confusions that last for *endlessly* long times.'

'Which words?' she asked.

'Any word that you *possess*,' said the excited spider doing some temporal weaving needlepoint. 'Possessive words like "mine," "yours," "his," "hers," are really confusing when you live in a world without any space in it. You can't separate things. My words are your words are his words. And in fact I am he, and he is we, and we are all you. You. You.'

As the spider got more excited, the word 'you' filled it with excitement and that set the web a-quiver! The words jumped up and hung themselves on the lines of the web.

Alice had to laugh. 'This is really so funny. Every time you say "you,"' she said, whispering the word 'you' to not shake the web, 'the words hang themselves on the web-lines. Perhaps that's where we get "remember your lines" when we talk about actors in plays and such,' Alice speculated.

Meanwhile the spider was singing better, crooning to itself and sounding like a cross between Rosemary Clooney, Frank Sinatra, and Kermit the Frog.

'You do something to me, something that simply mystifies me,' sang the crooning spider playing with its temporal weaving. As it sang, it quieted down, and then ended up hardly singing at all.

'Well I guess I better give you some more information. Now watch me as I weave my tale, or is it my tail?' as the spider jumped from one temporal line to another.

The spider then leaped from one side of her to the other. As it jumped, its words, 'now watch me' spouted under her feet, hanging from a web-thread the spider had just spun.

'See those words?' asked the talking temporal weaving spider. 'They are stretched out in time. That's why you can see them. Just before the "n" in "now" we have your immediate past and just after the "e" in "me" we have your immediate future. That's why we call my tracery a web of "timelines."'

Alice understood just what the spider was saying. Here, where there is no *space*, everywhere you look you see *time*. And when you look in one direction, you are looking towards the past, your own past, which is the past for some but not all. And when you look in the opposite direction, you are looking toward the future, your own future, and that holds for everyone who has a future—but not for those who are ahead already in the future looking back at us. 'But what about over there?' Alice asked. 'Is that side-wise in time? And what about up there, when is that?'

The spider was just about to answer when suddenly one of the voices that Alice heard earlier grew loud once again. It was the quirky professor once more.

'In this spaceless space, in this world of time without end, we are more than we seem and we are also a lot less,' the professor expounded loudly and then more quietly when he sensed someone was listening. 'Here we can't be separated from each other because there is no place to go—no place where we aren't. Where do you begin, and where do you end? Where do I start, and where do I end?' he asked.

Alice wondered to whom the professor was talking. She looked ahead and saw that he was talking to himself in the future, and she looked behind herself and saw he was talking to a young bright-looking kid that she recognized was the same professor when he was a teenager. She began to wonder, which one was the real Professor Flow, the one talking now, the kid to her left, or the old man to her right? Then she defocused her eyes so that everything became very blurry and she saw that the professor was every-where she looked! In fact, he was even *in* the spider and a *part* of the web at the same time! It was very weird to say the least, or for that matter, to say the most, because when she thought anything at all her thoughts began to hang out like wet clothes on the hairs of the professor's beard—which all of a sudden were exactly the same lengths as the timelines of the web itself!

'What was unbelievable in the beginning of the twentieth century has become humdrum as the twenty-first continues to unfold,' the professor went on. 'Back then even Einstein believed the new quantum mechanics—which entangled matter and mind—was one tiny rung up the ladder of credibility from Voodoo. But now we are beginning to see that mind and matter are very closely intertwined, very closely indeed.'

Alice began to 'see' just what the professor had in mind. All she had to do was 'look forward' and his words—the words he hadn't spoken yet—would be there. She was wondering how she could see words he hadn't yet said.

Then it dawned on her. If there is no space, and there is no separation, then all of us are the same. We are all one. Perhaps that's what all of those old mystics were talking about. Perhaps this was the 'spaceless space' where they go when they want to 'get away from it all.' Maybe Professor Flow wanted to make the soul more than just wish-fully wistful thinking; more than just a concept limited to the domain of spiritual practice—to make it accessible for a larger, more secular audience.

Alice looked ahead and saw and heard the professor's words.

'The soul, being both a part of the universe and containing the universe, is in fact, *the Universe*. We can no long separate everything that we thought we could separate. Things are interpenetrated in the world of spaceless space and all time. All was one. All is one. All will always be one. And when I think "I," the experience of that thought is universal to all who can think it.'

Just then the spider joined in and began to laugh so hard that the words fell off the line! 'Tell us professor how *do* you *do*?' asked the laughing still temporal weaving spider.

'I *do* fine thank you and how *do* you *do*?' responded Flow.

'You *do* something to me...' sang the singing temporal weaving spider.

The spider was singing again, Alice feared. She was afraid that if the spider sang too loudly the web would shake and then she most certainly wouldn't be able to see the rest of the professor's words ahead on the web-line, or worse, would fall into oblivion.

'Do *be do be do*. The palms are swaying. Do *be do be do*, is what I'm saying,' sang the laughing singing temporal weaving spider.

The web was gently swaying in the 'breezy tones' of the spider, but this time they were not shaking free, Alice realized.

'That's it, my friend.' It was the professor again. 'Either we are too busy *doing*, or not busy enough *being*, or we are too lazy *being*, and not *doing* enough.'

The laughing temporal weaving spider chimed in: 'We need to vibrate more. Get into the rhythm of it all. We need to dance, and sing, and...'

Alice was getting tired just listening to the spider. She wondered why she felt so tired. When she looked back along the web strand she was standing on, she saw herself as a young girl. 'But how can I be a young girl back there in my past when I am a young girl right here and now?' she wondered. Then she looked down at her hands and saw that her fingers were long and wrinkled!

Alice suddenly realized that she too was everywhere in the spaceless space of time and that she, too, was both old and young at the same time! She could, by simply focusing herself properly, jump from old to young in an instant. When she did, she instantly jumped along the timelines of the web like a schoolgirl skipping rope. Alice began to understand more fully just what the professor was talking about.

Then the spider started to dance and was drumming its belly with three of its arms.

'Wherever you go, there you are. The only problem with taking a vacation is you can't get away from yourself. Let me tell you about the time...' said the laughing drumming temporal weaving spider.

The spider, continually drumming its belly to keep time with its own monologue, was getting into the rhythm of its own shtick again and acting and sounding like a 'borsch-belt' comedian.

'Why, the spider is right', Alice said to herself, thinking of his words: 'wherever you go, there you are.' She really *was* able to be in two places at the same time. Actually, when she thought about it, she was really in one place, a no place (at that!), at two *different* times!

'But what does this all mean?' she wondered. She wanted the professor to explain to her what was happening.

'When the WIFF!-in-Poops assemble with their glasses raised on high,' sang the laughing drumming singing temporal spider.

'Of course,' Alice thought,' this is Wiff!-land! The web is a Wiff!-ian web of all the possibilities stretched out over time. The professor must have discovered something new to tell us. Perhaps he sees how the Wiff! is connected to consciousness and, could it be, to the soul? Perhaps the Wiff! is the Spirit?!'

Just then she heard herself in the future interviewing the undaunted professor. From the question being asked, her future self must have already come to the same realization.

'So the Spirit is the Wiff!' said the future Alice. 'And since the Wiff! dances and waves all the time, does this explain how Spirit becomes matter?'

The serious temporal weaving spider said: 'We should all reflect on this. Life is but a dream reflected to us like the Moon reflects the light of the Sun.'

The professor smiled at the spider. 'Indeed all is reflection, all is a kind of cosmic vanity. Not only does Spirit-Wiff! become matter, but it also becomes consciousness in a similar way. The Wiff!-ian Spirit reflects upon itself in two ways. Unreflected, the Spirit would not be conscious nor would it "Matter" at all. When it reflects in space it becomes matter, when it reflects in time it becomes consciousness.'

'Quite correct. Indeed nothing would "Matter" at all,' said the laughing temporal weaving spider, 'for nothing would ever come into being, no things would ever appear, and even space itself would not have even been thought of.'

'The Wiff!-Spirit would be both conscious and unconscious at the same time,' Flow went on, 'and it would be neither conscious nor unconscious, as well.'

Alice noted a kind of starry look appearing in the professor's eyes. Suddenly he was wearing a black beret like the San Francisco beat poets of the nineteen-sixties.

'Vibration happens,' rhapsodized the Poetically-Flowing-Flow. 'Vibration of Nothing. Vast vibrating Vacuum I am. Potential to be Anything. My Vacuum vibrates waves of potential. All me is powerful Desire to be. Nothing wants to be Something, yearns not to be Nothing. Everything in Creation seeks Value-Fulfillment.'

The spider was drumming louder and louder. Then Alice heard one of her Other Selves asking: 'Why is there desire to be *something* rather than nothing?'

The beat poet professor answered: 'Desire is the universal fundamental feeling. It is the drone tone. It is omnipresent. It is the vibration of spirit,' said Professor- In-The-Beat Flow.

The professor was snapping his fingers and appeared to be jiving. Alice ignored this. She wanted more answers now. 'How do we know this?'

'We know this based on observation,' answered the Professor-Scientifically-Speaking-Flow. 'We know this from models that accurately describe these observations. We know this based on the light we see from distant galaxies. We see an expanding universe. We have constructed a model of that expansion called the Friedmann model. The Universe will be 300,000 times bigger than it is now.'

When the professor said 'now,' the drumming suddenly stopped. All of the Other Alices looked confused and began to stare at each other, and then said unison: 'But when is now, now, now, now?'

The words began to grow everywhere on the web. Then the spider began to drum again and recite, causing the web to vibrate and shake the words free.

'How now brown cow?' said the drumming reciting temporal weaving spider. 'Now is the time for all good men... Now, now Mau, Mau. I want it now.'

Alice realized what was happening. Each person's point of view wanted its own 'now.' In fact the whole web was wanting to have its 'now' and 'eat' it too. But she thought it's not possible, unless...

Then the Professor-Scientifically-Speaking-Flow said: 'If we model the universe as a tiny fetus, then now it is—in comparison with what it will be—no larger than a germ cell compared to a human being. If you ask me: What the universe is expanding into? we find that there is no experiential answer. It is expanding into dimensions we cannot comprehend. Perhaps the fourth dimension, which may be related to time.'

Alice heard the professor's words. But she continued with her own thoughts, too. 'Unless the universe was expanding so that something new was continuing to appear, everything would repeat and repeat like the beat, beat, beat of the tom-tom.'

'We are fish in an ocean that we cannot ever see,' the professor continued. 'As nearly as I can determine, the soul began with a Big Bang when the universe began, and will end when the Big Crunch comes. From alpha to omega we have the soul. The soul is a Conscious-Wave-Form Self-Reflecting at the nodes of time where the timelines of the web cross. Consciousness requires a reflection. Matter is reflection of spirit at nodes of space.'

The spider was drumming again keeping time with the professor.

'Self is a reflection of soul in matter,' said Flow. 'Self is an illusion—actually a reflection of a reflection. It is both conscious and unconscious at the same time. We go into drama about ourselves, "I am ugly" or "beautiful." We are all in self-delusion. "I am 'I' the most..." "am 'I' the least." These are the delusions. These are the ego trips we all go on through time. Each trip is crazy. Each trip is based on the illusion of self. Self is mortal. Soul is immortal. Soul is reflected spirit. Soul is the verb. Spirit has to "soul" in order to become conscious. Nothing is unstable. The desire to create stable experience is the universal soul appearing.'

Then, in unison, the Other Alices asked: 'But why is it this way?'

'It is an unstable equilibrium. The nature of the universe is precarious. Matter un-cared-for may simply vanish. Stability requires attention. We shall overcome and evolve. We are evolving beyond what we can imagine today, but will imagine tomorrow. What the soul intends to do with all of its reflections only Everyone knows. But remember it's supposed to be fun.'

Then the professor and the spider popped like two soap bubbles and All of Time became suddenly silent.

Alone Alice surveyed the huge web. Then she felt another shiver. 'Curious,' she said, 'I'm shivering again and yet I'm not cold, not a single bit.' Then the shiver occurred again and she saw herself to her right. Then another shiver and she saw herself to her left. And then with the next quiver she saw herself expanding and filling all of the immense space she was in.

'Oh, dear, I'm really getting an inflated ego,' she cried out. When she realized what she had said, she laughed, and her laughter filled the universe.

18 ALLUREMENT

Nick Herbert
and William Brandon Shanley

In which Alice learns the Attraction is what Binds the Universe Together, Inside and Out.

Alice dreamed that she was in the command center of a futuristic spaceship, similar to the one used on the television series, *Star Trek*. She was talking to an older, scholarly looking scientist, Professor Collins. Both were dressed in slightly garish outer-space uniforms. Professor Collins had long gray hair, and a wide and mild face.

'This is your last voyage, Alice,' the Professor said. 'You've seen and experienced a great many things, but tonight we will draw all of it together as we visit the final frontier, the universe as a whole, the cosmos with all its vastness.'

He touched a button, and a huge observation screen slid open to reveal the millions of stars through which they were moving.

'What is the nature of the large-scale universe?' he asked. 'How does the human being fit into the universe? What can we possibly matter in the midst of such enormity?'

'I can't help but feel a bit dwarfed by such questions,' Alice said somewhat meekly as she peered out into the infinity of space.

'You have a lot of company in that,' he tried to reassure her. He began to walk and gesture to the cosmos as he spoke. 'From the beginning of time, humans have gazed upon similar sights and have asked themselves these same deep questions. The majesty of the night sky demands such reflection. Out of just such a vista as this the most ancient philosophic wisdom was born. It might help us to remember that as we reflect on the universe as a whole we are joining a very ancient tradition. All our ancestors wondered in the same way about the universe. We continue that endeavor tonight. But there is a difference. For we bring to the task a vast amount of information they did not possess. For instance...'

The Professor touched the control panel again, Alice braced herself, the ship zoomed off at hyperspeed, and focused on a single star.

'...we now know that our Sun is but one star in the two hundred billion of the Milky Way Galaxy,' he continued. The ship banked and soared through the countless stars of the Milky Way, and on into intergalactic space.

'All these stars rotate within the galaxy; our own Sun takes two hundred million years to make a single journey around the galactic center. And as we move outside our galaxy, we begin to see that the large-scale structure of the universe as a whole consists of some fifty billion of these galaxies, each one a dazzling jewel strung on an invisible net that stretches out and fills up all space.'

'It's just so overwhelming!' exclaimed Alice, visibly moved by the spectacle of light and the Professor's lecture.

'Alice, when you take in the view of the universe with all these billions of stars and galaxies all swirling about one another, is there any question that arises in your mind?'

'Well, I can't help wondering, what holds it together?'

'What holds it together? The answer is, "gravity." But "gravity," in one sense, is just a word. So we need to think beyond the word "gravity" to the reality it points to, a power that suffuses the entire universe and holds all of these fifty billion galaxies together.'

'But how does it do it?' asked Alice. 'How does this power actually hold galaxies together?'

'I'm going to tell you something that you will find hard to accept,' the Professor continued. 'But I'm speaking now as a physicist and I'm summarizing centuries of thought by some of the finest minds in the history of humanity. The way gravity holds the universe together—is by holding it together. That's the best that can be said.'

'What kind of an explanation is that?'

'A scientific one. It differs from earlier explanations in that it doesn't pretend to do away with the mystery of gravitational bonding. Isaac Newton was the first one to recognize this difference.'

'I really don't understand.'

'Everyone else was trying to give a reasonable explanation. The most famous was that of the medieval world. They theorized that the stars were stuck on these crystalline spheres. The reason the stars moved was because these spheres moved. This is what we would call a reasonable explanation, for it makes sense to us. We can picture it and we can feel as if we really understand what's going on. But in truth, there are no crystalline spheres out there, so the explanation is no explanation at all.'

'Please wait a minute! I'm just getting more and more confused. So what can we say, if we can't have any of these reasonable explanations of gravity?'

'The most basic fact of the large-scale universe is a pervasive power that holds everything together. The nature of the universe is this attracting power. That it fills the universe top to bottom and horizon to horizon.'

'And this is gravity?'

'Yes, but let's call it something different so we can escape the illusion that we somehow understand it fully. Let's just call it "Allurement," by which we can hope to indicate not just the gravitational attraction, but all powers of attraction throughout the universe. So, the reason the Earth revolves around the Sun, and the reason the Sun swings through the stars of the galaxy, and the reason the Milky Way Galaxy revolves around the Andromeda Galaxy is the same reason. It is because this attracting power of Allurement pervades the universe as a whole, and draws all things into bonded relationships.'

'And that's the most we can say?'

'We can say many more things, deeply intelligent things, about the dynamics of Allurement. But the most basic fact, beyond which nothing can be said, is simply that the nature of the universe as a whole is this power of Allurement that draws all things together. Just look at the night sky.'

He indicated the observation screen with the stars and the galaxies.

'In order to see what is before us, we must understand that we are not simply viewing a great conglomeration of things. There is more here. All of these stars are held together by an invisible, alluring power. Each one is directly bonded to all the rest. That's the fact of the matter. To reflect on the universe as a whole we need always remember this power that sustains the universe's order.'

'So Allurement is what holds the universe together?'

'Yes. Just imagine what would happen if you could find some cosmic switch and turn off this invisible power of Allurement. First off, all the stars of the Milky Way Galaxy would no longer hold themselves together, but would disperse in all directions. Our Sun would drift off into space, disconnected from everything else, and then it really gets bad. The hydrogen atoms of our Sun would no longer attract one another, and would also fly apart, thus extinguishing the nuclear fires of the core star. All the stars would lose their light. You see? Take away this invisible power of Allurement and the universe dissolves into dark dust.'

'Now I can really see why it's important to remember Allurement when we look at the night sky!'

'And we haven't come to the most important activity of Allurement, which is to draw forth all creativity. All creative activity is evoked, or initiated, by Allurement. The easiest way to see this is to picture a great cloud of hydrogen atoms floating in the middle of the Milky Way Galaxy. Each of these atoms is drawn toward all the rest. By giving in to this attraction, the hydrogen atoms rush ever faster together into the implosion of a new star. You see?'

'I think so.'

'Allurement is what starts the creativity that results in the birth of the star. If the hydrogen atoms did not experience this power of Allurement they would stay floating in their cloud. But Allurement draws them out. By pursuing this attraction, they find their way into their destiny as a star.'

'And all this creativity is the same, throughout the universe? Even here on Earth?'

'Exactly the same. It was only because Earth experienced its Allurement to the Sun that it entered into a five billion year relationship that resulted in the birth of life and human intelligence. We are, all of us here on Earth, the direct result of the attraction that brought Earth and Sun into a stable relationship.'

'So even though we don't understand the nature of Allurement completely—'

'We'll never understand Allurement completely. Do you think hydrogen atoms had any chance of understanding what drew them out? Of course not. But by pursuing that Allurement they gave birth to the stars.'

'And so for humans—'

'We were created out of Allurement, and we find our own destinies in response to the Allurement that we experience—'

'Like the stars—'

'Beset by deep attractive powers—'

'And so we too—'

'Are pervaded by fascinations, by dreams of what might be, by alluring and haunting possibilities. It is only by pursuing these, each of them so uniquely personal, that we find our true destiny.'

'In response to Allurement, to find our way forward—'

'And to become Allurement itself. To become that which initiates and evokes creativity. We are drawn forth by the alluring power of the universe so that we might find our destiny not apart from, but *as*, Allurement, as the very power that evokes the creation of the universe.'

'Just like the stars.'

'From which we came.'

'To awaken to Allurement—'

'To become Allurement itself.'

'So the Allurement between atoms and molecules the Allurement between stars, the Allurement between people, isn't that what we call *love*?' asked Alice.

'Well,' the professor paused, 'I guess you could say that. Yes. But the word love is a problem because people think only of human love.'

'But if we take a broader view, couldn't we say this force of love is what binds everything together in a creative embrace? Isn't that the Grand Unified Field that Dr. Flow's been searching for?'

'Clearly, Alice, the strong and weak nuclear forces, electromagnetism, gravity and human love are all very different things...'

'But isn't our seeing them as different things a result of reductionistic thinking? Our worldview? Our minds? Our consciousness?'

'Er, yes, you could say that...'

'And, isn't it we who differentiate them in our laboratories for our own purposes?'

'Yes, Alice...'

'So, what if we called all these different forces of Allurement "love"? What if we say that the vibration of the universe is love? Wouldn't we all be *in* love?'

'Then, it seems, we'd all be completely in love, Alice. All that there is, is love.'

'That's it! Born in love. Attracted in love. Searching in love. Creating in love. *Being in love. Living in Love.*'

'This is what poets and mystics have been telling us for ages.'

'This is what lovers have always known.'

'Being in love. And the light...'

'I am love, I am the light. I want to go into it...'

'And so do I...'

'For now, and forever.'

Alice and Professor Collins' eyes locked together. There was applause from the flight crew. Someone popped open a champagne bottle and distributed glasses all around.

Then something very strange happened. The Professor started pulling on the skin of his face, making a horrible expression, startling Alice. But as she continued, Alice noticed as he continued to pull at his face, he was actually removing a rubbery, latex mask. And wow! In an instant he revealed himself to be tall, dark and handsome twenty-something with shiny black hair!

'You're not Professor Collins! Who are you?' Alice exclaimed.

'That's right, Alice. I'm not Professor Collins. Surprise, surprise, my name is Allure!'

In a blink without a referent, Alice and Allure were already toasting one another through intertwined arms. Alice was especially animated as she let go, falling into his deep, dark eyes. Behind them, the large-scale universe seemed to respond with its own unique pyrotechnical display of cosmic appreciation.

19 THE SYMPHONY OF CREATION

F. David Peat

In which, in the Company of an Odd Trio on the Beach, Alice experiences the Universe as Non-Unitary Musical Invention.

Alice felt that her journeys were coming to an end but there was one place she wanted to see again, a place where she had been happy, San Francisco. And so she blinked her eyes and opened them again to see the vendors' stalls outside Pier 9. It was a warm summer weekend and Alice intended to spend the rest of afternoon wandering along the boardwalk, looking at the stalls and staring up into the sky at the kites flying in the breeze. But after a short while she began to feel hot and tired.

'I'll just sit down on the grass for a moment,' she said to herself, 'and read my book.'

But it was a difficult book and, since it had no pictures and very little conversation, Alice found her attention wandering to the people who were passing by. Facing the sea, a man was dancing and waving his hands in the air as if conducting an imaginary orchestra. He was dressed in old-fashioned clothes and wore a white wig so that he looked a little like God. Near to him an older man was riding a penny-farthing bicycle and towing a large mass of scrap metal. But what attracted her most was an elegant middle-aged lady in a long dress who kept looking out to sea, and then bending over to write in a notebook that rested on a wooden board on her knees.

'How very curious people can be,' mused Alice as she stifled a yawn, for by now she had become quite tired and decided her book would serve better as a pillow than as something to read. And so she shut her eyes, but only for a moment she told herself, and listened to the sounds which surrounded her. From where she lay on the grass she could hear the noise of the traffic and the shouts of the street vendors. From farther away came the laughter and cries of children playing and behind it all the soft, sleepy sound of the waves. But soon the sounds seemed to fade and for a time she could hear nothing, as if she were floating away on the pampered comfort of cotton wool.

'Maybe I'm inside a cloud,' Alice thought. 'If I am, then it's very pleasant. Much nicer than in an aircraft.'

Again the voices of the children blew towards her on the breeze and she imagined that they were all playing in a nursery together or running through a large garden, like the sort of garden she used to read about in her storybooks. One of them had found a spider's web, and another saw a great metal ring flashing in the sunlight, while a rather nasty boy kept shouting about a great beast that stamped its foot. Alice wondered if he was referring to a giant's heartbeat or simply to the sounds of the waves. And then her head was filled with the sounds of church bells so that through the cotton wool of her dreams she wondered if she'd floated into a picture postcard world and would open her eyes and find herself sitting under a willow tree in one of those English meadows with a beautiful village and an old church nearby.

She was about to drift off again when a rather posh voice awakened her.

'You've been asleep long enough, haven't you, my dear?'

Alice opened her eyes to find an elegant lady standing over her and smiling. 'Oh, I'm sorry. Were you speaking to me?'

'I was watching you while you slept. You remind me so much of a photograph I saw long, long ago—a young girl sleeping.'

'And was she like me?'

'Exactly. Her name was Alice like your own. Alice Liddell.'

'But my name *is* Alice. How did you know that?'

'My name is Mrs. Woolf.'

'Oh, like the wolf in the fairy story.'

'Certainly not, but I do happen to know a great number of things. Now, let's get down to brass tacks and tell me about your dream.'

'I wasn't really dreaming. Just listening, listening to the children and the sound of the bells.'

'And I've been listening to the waves. There is the difference between us,' said the elegant lady in a rather annoying way.

'But the bells are really quite nice. They go on forever and ever and the pattern never repeats. I think that's what time is like. And I know all about time because I keep it when I play the piano, and sometimes I even have to beat it,' said Alice proudly.

'And the cotton wool?'

Alice looked at her with a start, 'How did you know about that?'

'Which are more real, Alice, those moments when you are aware of everything around you? Or those cotton wool times when you seem to be floating away and dreaming about the universe? Or does the universe happen to be dreaming about you?'

The conversation had taken such a strange twist that Alice had to pinch herself quite hard to find out if she was still awake.

'I thought it was only scientists who thought about things like that,' she said. 'All the ones I've met up to now are always asking me questions that end up making my poor head ache.'

'But, my dear, poets, artists, composers and writers have been asking questions long before scientists were even thought of. Questions such as: What lies behind the reality when we shut our eyes? And where does creativity come from? Or dreams? But let's get back to something a little more concrete. Those bells you've been listening to, and my waves. Now tell me; do bells really go on changing forever?'

'Well, they certainly do sound different,' said Alice.

'But that's not the point. Does the sound really change or is it simply a matter of what scientists call permutations, of people ringing the same set of bells but in a different order?'

'Does it make a difference?' asked Alice.

Another voice boomed: 'It certainly does in this truly most marvelous universe of ours.' This was followed by a tremendous crash as penny-farthing, scrap iron and rider came to a head-over-heels stop in front of Alice.

'Allow me to introduce myself,' said the rider, taking off his top hat. 'I am The Wheeler. And this is my most wonderful, stupendous, fantastic, supercalafragalistic machine,' he said, pointing to the mass of scrap iron. 'I bet you don't know what it does.'

Alice, who had been exposed to too many surprises by now, was certainly not going to be fazed. 'Oh, I suppose it's just some old time machine or something.'

The Wheeler looked rather taken aback. 'Well, to be strictly accurate it's a marvelous space-time machine. But at the moment it's not all that great at generating ripping, up-to-date, genuine, 100 percent good old American time…if you see what I mean. But it's really not at all bad at space-time—provided you don't look too hard at the time part of it. Watch!'

The Wheeler took a wrench and screwdriver from his coat and began to bolt the various pieces together until pretty soon Alice was looking at something that looked a little like a barrel organ with a large funnel at one end and a knitting machine at the other.

By now The Wheeler had grown very excited. 'You put the raw material in there and brand new space-time comes out of that end there.'

The Wheeler took down a dirty old sack off the back of his penny-farthing and tipped the contents into the funnel.

'But they're just lots of rusty old nuts and bolts,' Alice protested.

'They may look like junk to you, but to me they're the nuts and bolts of the universe, little girl. Take a closer look.'

Although they were indeed rather rusty, Alice saw that each was covered with curious wiggly symbols. As they warmed in the palm of her hand she noticed that, like little magnets, they began moving toward each other. And when they touched they locked together so perfectly that she could no longer see any join.

'What you're holding in your hand, Alice, is pre-space. It's made out of the most amazing and purely logical and most abstract relationships— the primary connections of those things that come before our thought and consciousness. Logic was present before mathematics and before the laws of physics. It's the origin of everything. Just watch what happens when they get working inside my machine.'

The Wheeler pointed to the other end of the machine where the knitting needles were hard at work clicking away and producing a misty foamy sort of stuff that glistened and sparkled until it seemed to dissolve away like steam from the mouth of a kettle.

'Look, Alice, it's making space-time,' The Wheeler cried. 'It's making the food of the universe, it's knitting the stuff that everything's made from— stars, planets, and even little girls. It's the basis of matter and energy, and all the forces that push and pull you around. The backdrop of quantum theory is nothing but space-time.'

The Wheeler rubbed his hands and danced around until, with a great puff of dirty, black smoke the machine came to a sudden stop and a lump of metal fell with a dull clank onto the ground.

'That's the problem,' The Wheeler said, looking somewhat glum. 'It keeps breaking down. If it was a really marvelous, good cosmos knitting machine, then I wouldn't have to keep fixing it. In fact, it would have already knitted itself anew and the whole thing would begin to fly away all by itself without the need for me to switch it on and keep it warm and feed it paradoxes every night. There's something fundamental missing. It simply refuses to come alive. It's never going to create a new universe all on its own.'

'For that you need art,' said Mrs. Woolf. 'It's like those "Bells of Alice," they simply won't come alive by themselves. All they can do is follow the rules one creates for them, they never generate anything totally new, or at all musically surprising.'

Alice looked up to find that the dancing man in the white wig had joined them and was nodding vigorously in agreement.

'Johann Sebastian,' he said, holding out his hand for a moment. 'Those bells, Alice. I really must have a word with you about them.'

The man in the white wig explained to her that bell ringers produced what are called 'changes'.

'Suppose there are five bells in a belfry,' he said. 'Well, each time they ring the bells in a different sequence ABCDE, BACDE, ADBCE, ADCEB, and so on until after 120 changes you're back at the beginning again. The more bells you have in the belfry the longer it takes before you ever repeat the same sequence. With twelve bells it would take 479,001,600 changes, and so all the bell ringers would be long dead before they'd hardly even started.'

'All that happens with your bells, Alice,' joined in The Wheeler, 'is that the same things simply get reordered. Start with a given sequence of bells, agree on a rule to permute them, and you can predict exactly what things will sound like in an hour from now, or a day from now. It's the same with my poor old knitting machine. It can only create what scientists call a "Unitary Universe." The past exactly predicts the present and, given the present, you know exactly what the future is going to look like. It's true that things get shuffled about a bit so they all look different, but the whole thing is just stirring the pudding. Nothing really new is ever going to be created. My poor old cosmos knitting machine is simply going to churn out the same unitary universe day after day.'

'Yes, Alice,' Johann Sebastian added. 'You may have thought that your bells were like the ticking of time on a clock. But that's not the same as living, dynamical time. It's not the time that powers the universe from its moment of creation. And it's certainly not the time that's present in my music.'

'Art is the missing ingredient,' said Mrs. Woolf, 'Art and Nature. For both are alive, and creative, and both allow the new to be born. The Wheeler's cosmos knitting machine is a little like a kaleidoscope that makes variations on the same pattern. It can never break outside the rules that have created it.'

'Music has its rules and Nature has rules, just like The Wheeler's knitting machine,' exclaimed the man in the white wig, getting very excited. 'Everything has rules, but in Nature and music something new is always being created from inside the rules. And the rules themselves, the symmetries, the orders, the patterns are themselves subject to the creativity of time. Isn't that correct Virginia?'

Mrs. Woolf nodded: 'And on every occasion I listen to your music, it changes, because I'm bringing something new to it. Rules are just rules, but Art and Nature depend on contexts, and contexts are always changing.'

Alice chimed in: 'Yes, I remember seeing a movie a second time years later and I noticed that I perceived new levels of meaning because my awareness had grown with the years.'

'Yes, yes, yes,' said The Wheeler. 'Remember how in quantum theory the observer and the observed are always a part of one whole? Well, at the same time everything is in flux, everything is ever changing. The deepest logic that ties our marvelous universe together can never be static, its context is always changing.'

The Wheeler grew rather sad. 'That's the problem with my cosmos knitting machine. It's a unitary machine. It constantly operates according to the same rules. What I need is a non-unitary knitting machine. One in which something new is always around the corner; something inspiring, novel or surprising. If I could build one of those, then it would take off and fly. It would knit itself anew, and create another wonderful, marvelous, surprising universe.'

But by now Alice was so puzzled that she had to sit down again and take her head in her hands.

'But all this began with my bells and your waves!' he remarked to the elegant lady. 'So what's so special about waves?'

'The sound of the waves is the rhythm of art and all of life on this planet. Waves have their ebbs and flows, their repeating rhythms, yet at the same time they are always new. There is always the possibility of a sudden storm, or an extraordinary high tide.'

The Wheeler nodded. 'Scientists call the waves an "open system." You can't draw a boundary around waves or imprison their sound. The waves beat on the shore because of the Moon's gravitational attraction on the sea which causes an alteration of high and low tides. But there's also the effect of the wind, which changes daily and hourly because it's produced by the movement of weather fronts. And the great motor that powers this weather is a combination of the Earth's rotation and the Sun's heating of the Earth's atmosphere. But things don't even end there, because Earth, Sun and Moon are all rotating, and the sun's energy output is constantly fluctuating. The entire system ends up being so big, so complex, and so sensitive that you can't ever draw a boundary around it, or describe it by a single set of rules. It is a totally free, "Non-Unitary System": one in which the new is always waiting to be born.'

'That's why the rhythms of the waves, and of all Nature are like my music,' said Johann Sebastian, as sounds of brilliant harmonic complexities suddenly actualized their thoughts, lifting their hearts and spirits to the sky. 'Give me a theme and I'll write you a suite, a fugue, a set of variations, anything you like. The music flows from that original theme, according to a set of rules, just like the patterns of sound flow from the bells. But art is different from mathematics because a composer can constantly bend and shape and transcend the rule of his composition. I can introduce surprises, twists and turns that you would never expect. If I left a fugue unfinished, then a clever student may deduce how I would end it, but he can never be sure that I will not introduce a new twist to the theme, or try out an unexpected modulation.'

'If we must use The Wheeler's dreadful terminology,' Mrs. Woolf added, 'then both Art and Nature are "Non-Unitary." Examine a piece of music, a poem, a novel, or look at the sketch for a painting, and you get an approximate idea of where it's going. But that future is never totally determined by the present. There will always be room for something new, something unexpected, a thing that is totally fresh. And every time you go back to that creative piece, you yourself will be new and fresh, Alice, and so you too, will add something new to it.'

By now Johann Sebastian was dancing from foot to foot and waving his hands in the air. 'And this music goes far beyond anything I could write, or The Wheeler's machine could ever knit. It's the "Symphony of Creation." It's the music of all of Nature—running streams and the joy of bird song, the growth of grass and the waves beating on the shore.'

By now The Wheeler had begun to dance and his voice joined that of Johann Sebastian, so that Alice could no longer tell who or what speaking. 'All of Nature is constantly forming into patterns. Form and structures are everywhere. That is how we recognize the world, in terms of its symmetries, repetitions, patterns and forms. Everything from the structure of a fugue and the pattern of a sonnet, to the spirals in a sunflower and the curve of a snail's shell, "The Universe is a Work of Art."'

'And patterns break up as well, water boils, storms whip up the surface of the sea, machines rust, animals die. At every moment forms and patterns and stabilities are emerging out of the flux and chaos of the world, and dying back into it.'

Mrs. Woolf, being somewhat reserved, had not yet taken part in the dance, but her deep voice had begun to join the others. 'In our moments of total awareness we see the patterns of the world like the facets of a crystal, hard and clear and shining. But there are also those moments of cotton wool when we dissolve back into a padded insensibility, and the world within.'

'Now I understand what you are saying!' exclaimed Alice, dancing in excitement. 'Our minds and the whole universe are reflections of each other. No, they are one and the same thing. We create patterns and moments of awareness in the world and then we dissolve back into some sort of great pattern of which we are also a part, but it has become so complex that we can no longer sense our identity within it.'

'And in those moments of cotton wool each one of us becomes a note within the great music of the universe,' Mrs. Woolf added, sounding quite happy, 'in the "Symphony of Creation." It is only in those instances when we separate ourselves and things become clear for a fraction of a second that we can see the music, the patterns, the cosmos as something objective and outside ourselves.'

'The cosmos is the great heart beating. It's constantly coming in and out of existence,' cried Alice. 'And that's what the boy meant when he spoke about the great beast with its foot, stamping.'

'It is the breathing of an enormous organ,' sang Johann Sebastian, 'an organ that stretches throughout the universe playing an endless fugue in which there are ever new themes, new transpositions and new variations. The fugue goes on and on and each one of us, and every atom and electron, is both the player and the played.'

By now the dancing had become so joyful that Alice forgot how to listen, because the music seemed to be emerging out of her own body, as if every part of her was singing. And for a moment the phrase 'music of the spheres' came into her mind, but then she was dancing, and flying through the sky, and falling through the cotton-wool clouds and all around her were the sounds of bells ringing, and birds singing, and the waves breaking on the shore. She could feel the atoms joining in the music, and electrons dancing like colored lights until, suddenly, they exploded like great fireworks into cascades of tiny, bright, vibrating strings.

And as the strings spun and swirled into the velvet blackness, Alice realized that the strings were not only inside her body but that she too had become stretched out until she was a single vibrating string that extended across the whole universe, so that all the planets and stars and galaxies were contained within her. And then time ceased and there was only a warm comforting darkness, yet a darkness filled with a music that seemed to be simultaneously present in all its notes and themes and movements. And Alice realized that this music had always been there but she had never really stopped to listen before.

And a little later the music began to fade, and when she opened her eyes she was back at Pier 9, and The Wheeler had cycled away, and the dancing man was nowhere to be seen. And when she looked for the elegant lady she could just make out, in the far distance, someone walking in a large floppy hat; someone walking slowly towards the sea and the waves.

Alice felt so tired that she closed her eyes. And pretty soon she was deeply asleep and in the warm summer evening began to dream a very strange dream.

20 ALICE RIDES THE PHOTON

William Brandon Shanley

In which Alice takes a wild ride with a Photon through Creation–yet doesn't go anywhere at all!

'Light falls in love to create life,' was the thought Alice heard when she awoke from a lucid dream dancing with photons.

Disoriented and squinting into the bright light, Alice felt reassured when she heard the warm, familiar voice of the photon again coming from everywhere.

'That's right, I'm back,' said the photon, 'but as you may have gathered, I never left you.'

Alice was taken aback. She hadn't heard from the photon in quite a while.

'How can that be?' Alice asked, still trying to adjust to the light.

'I'm a bit of an actor, I must admit, Alice, and one of my acts is you.'

Alice was just beginning to feel insulted by the pretentious assumption, when the photon interrupted.

'Alice, remember when I said that I am action but from your perspective, I don't travel anywhere at all?'

'Yes, I remember trying to visualize what you were saying in a thoughtful, logical way, but didn't have much success,' said Alice, feeling a little queasy with the onset of yet another paradox. 'What does that have to do with me being one of your acts. I am me and you are you.'

'Alice, do you remember, I said I was qualitatively different from other quanta?'

'Yes,' she said, 'and I also remember thinking how you were being quite boastful and persistently pushy.'

'I will admit being very proud of my acts, Alice, for besides being you, I am all the others, too.'

'Now please, Mr. Photon, besides being nonsensical, don't you think you're getting too personal? I listened and went along with your story before, but now you're confusing me again,' Alice lamented, scratching her head. 'A moment ago you said you were qualitatively different from the others, now you're saying they are really you? That they're your acts—and I am, too?'

'That's right, Alice,' assured the photon. 'Now hold on a minute and let me explain. Do you remember we discussed Mr. Einstein?'

Alice nodded.

'Dear Albert proved to the world that matter is condensed energy. Well, I'm here to tell you I am the source of it all.'

Alice was not buying it.

'You're being rather rude again by making such egocentric claims,' Alice huffed.

'Patience, my dear. You'll understand me better in a minute,' the photon offered. 'And do you remember how I said I have no mass, and from my perspective, I don't go anywhere, or experience space and time?'

'Yes.'

'Now try to understand what I'm about to say. I have no mass and no space and time, and from your perspective, when I slow down, you observe mass, distance and time coming into existence, and in that instant, all physics is born! Continuing my descent, I give up energy to form quanta: electrons, neutrons, protons—and the possibility of the entire colorful, charming, quarky zoo ad infinitum.'

Alice was trying to take it all in and following just the words because she felt that this time, more than ever, she was losing her grip on reality.

'You see, physics consists of three parameters: mass, space and time. I am the source of them all.'

'Oh, I remember what Mr. Einstein said about matter being condensed energy quanta. And now you are telling me that all the other quanta come from you. Is that right?'

'Precisely.'

'So from my perspective, your actions produce all the quanta in the universe?'

'You're catching on beautifully!' said the photon with glee. 'Now you're beginning to understand my act.'

'I see,' said Alice, as her mind percolated with all sorts of images and her body felt a thrill and shuddered. 'But you also said you were action. How does that come into play?'

'Here's how. I am a very special action. I become all things, all at once, in every possible variation. Here's how I do it,' continued the photon. 'From your perspective, as I continue my descent and condense into quanta, I'm giving up energy to transform myself from a state of action, total freedom and pure possibility into everything you experienced in Quantumland. This state of freedom is what scientists call uncertainty, but that's only how it's viewed from their perspective of control. I am actually unlimited possibility and freedom to become everything in the universe. As I continue my descent into matter, notice how my quanta bond into just the right number of atoms, and my energy bonds atoms into molecules. Quite an act, right?'

'And a tough one to follow!' Alice joked. 'But what do you mean by "just the right number of atoms?"'

'To create life,' Alice.

'But wait a minute,' Alice interrupted. 'So what you're claiming is that you are action and your goal is to manifest everything we experience in life, our universe, even life itself? Is that it?'

'Quite right, my dear,' the photon confirmed. 'My life is an action that creates all the fundamentals to manifest higher and higher degrees of perfection. In your world, my best act so far is being human. And even though it's a work in progress, a quite exquisite one it is: ten to the eleventh power brain cells capable of interactions greater than the number of atoms in the universe. Trillions of cells of the body with faster-than-light connections, perfectly choreographed in an exquisite dance of harmony and balance. A multidimensional Mind interconnected and interpenetrated with everything in Creation's web of love.'

Alice was captivated by what the photon was saying about the dance when she tuned back into what the photon was saying.

'But how is that possible to create life from molecules, Mr. Photon?'

'Molecules don't create life. I do. You see, Alice, my purposive action to become, my will, if you will, is the fourth parameter that Newtonian physics, Darwinism and other mechanical models exclude from their theories. This is consciousness, the life force, the *élan vital*, bioenhanced energy, metamorphosis, the drive of evolution, the Mind, if you will, that you've been hearing about in Quantumland,' the photon informed her.

'OK, I think I'm following you so far,' Alice said.

'A friend of mine named Arthur M. Young surmised my true nature and wrote about my "arc of life." You see, Alice, Mr. Young was an inventor of the helicopter and realized in making calculations for controlling flight in three dimensions, something other than mass, space and time was required for acceleration. It may seem obvious to us now, but the missing parameter is control. Control is what your father exerts when he chooses to push down on the gas pedal to accelerate the car, and vice versa. In so doing, he is making a choice about how to move through time and space. This choice is called "will." It is also called "freedom."'

Alice had a question.

'But you said as you fell into matter you were giving up your freedom. Aren't you contradicting yourself? And what does the fourth parameter have to do with creating life?'

'From your perspective I'm giving up freedom, but I'm really transforming into states that make life possible. I'm expressing my Quantum Potential, you might say! It's like this. My descent, my fall, my entropy ceases when I bond into fairly fixed and determined molecules. Next, instead of continuing to give up energy, I begin to take energy from the environment and perform my most significant act on my journey to create life. I do a U-turn, become neg-entropic, launch myself up from the fixed molecular platform, use my purposive action to create polymers, then cells, plants, next animals, and humans, each with a higher degree of freedom than its predecessor, each with a higher degree of complexity and perfection than what preceded it, each including within it what came before it.'

'Why you're taking my breath away, Mr. Photon,' Alice swooned. 'The implications are astonishing, yet they way you describe things, you make all the complexity sound so simple.' She paused for a moment. Then, 'And so what happens next? What's your next act?'

'Complete freedom. I return to my original state,' said the photon. 'And that's my story, my whole act. Again, Mr. Young calls it my "life arc." It's has the same trajectory and character arc as universal stories of the Fall and Resurrection of Man. So all those stories are my story, too.'

Alice thought for a minute.

'But where do you come from? Who births you?'

The photon paused for a moment and then said quietly, 'I come from the Source of the Totality, the Ineffable Beyond-Stuff some have called God, Yahweh, Allah, Creation, All That Is, the Eternal Sponsor.'

The synapses in Alice's brain exploded like a binary star, putting down whole new dendrites and cross-circuiting axon pathways. She nearly fainted from the illumination, and felt her body thrill and chill. In an instant, she was beyond her body, in a state of pure consciousness, freedom and action. She noticed she became whatever she thought. She thought about Quantumland and yearned to be with her amazing friends again.

All at once, Alice became a superposition of all the zany experiences, fascinating characters and quantumstuff she'd met in Quantumland, Chaosland, Biologyland: Professor Flow, the Wiff!, Wavicle, Professor Who, Drs. Yes & No, the Quantum Hussy, Schrödinger, the Cheshire Cat, Queen Rosie, Goswami, Yellow Cab, Beatrice the Cook, Myles na gCopaleen, Kat's Sfonk, the Little Biologist, Hamilton, Berkeley, Ireland, New York, San Francisco, Sggirb, Dr. Goswami, Luc the Caterpillar, the Wheeler, Mrs. Woolf, Johann Sebastian Bach, and cotton wool. Bells, zaapps, giggling into chiming delight. She danced a jig with Mr. Collins, the Time Weaving Spider, and Allure. She morphed into the Copenhagen Interpretation, mind over matter, a hologram, Bohm's neorealism, Everett's parallel universes right on top of Darwinism, DNA, Heisenberg's Duplex World...and...

The photon.

In that no time, no place NOW, Alice was at home, realized who she was, experienced being at One for Once and for All Time with a Sense of Meaning and Purpose to Become All She Could Be. To be free at last to let Alice be Alice.

'Oh, what a marvelous trip it's been, is and isn't,' Alice giggled across Creation and to whole cast joined in a spectacular, multidimensional dance of light, love and life.

Then came that wild, now familiar surmise, 'Light falls in love to create life.'

Thank you for the dream, Lewis Carroll!

ALICIAN SCIENCE

Jean Houston

The science of Alice starts as nonsense and logical paradox, but looks forward to many of the great scientific speculations of the 20th and 21st centuries. Each period of paradigm crisis sees the emergence of the next shift out of renewed interest in the odd, the esoteric, the occult, or in a refreshment of nonsense. Deeper strata of the unconscious are called into play.

For example, the late antique world turned to occultism out of an inability to carry Greek science any further forward. Alchemy arose out of a similar retreat from the stuffiness of medieval scholasticism. The ideal of the magus in the Renaissance itself stimulated research into the nature of things and brought about the emergence of mechanistic science. Leonardo da Vinci was stimulated by Hermetic heresies. Translations of the Hermetic books by the court of the de Medici family and Ficino were a catalyst for deeper thinking about the nature of the world with its consequent discoveries.

The 19th century saw perhaps the greatest change in the state of being alive that the world had ever seen. Old beliefs and ways of being were threatened on a major scale and this ambivalence showed up not only as a collective breakdown of mind—but an honoring of eccentricity, a rampant hysteria, a weird eroticism, just under the surface crust of niceties of the Victorian era.

Thus, Charles Lutwidge Dodgson, an eccentric bachelor lecturer in mathematics at Oxford, is symptomatic of this era, and of the fact that what appear to be 'screwball' ideas will emerge in state-changing times as premature manifestations of a new paradigm which are similar to the initial bubbles in a fluid as it comes to a boil. Such a change in state will generate chaotic imbalances that materialize as perturbations in the social and intellectual fabric. Most of these will simply be aberrant curiosities which cannot be located until a context has emerged, as more bubbles become evident. Such perturbations eventually take on their own interactive patterns out of which a new paradigm is formed. Now, as with some of the 'screwball' ideas in the two Alice books—so many of these controversial ideas will simply dissipate, particularly if they emerge too early.

The constructs rise from a deeper order of Mind that is getting ready to emerge. Such theoretical constructs—like the weird science of Dodgson's alter ego Lewis Carroll—run the risk of appearing psychotic since they pose answers to questions that have not yet been asked, much less consciously formulated. The scientists or poets or mystics who pose these premature questions are called 'crazy.'

Lewis Carroll, that shy and stammering genius, was called 'delightful' because he presented it, perhaps even unknowing to himself, in such a way that it could reach millions of children and their parents in a 'nonsense' fairytale—and help seed the dream of the 20th and 21st centuries. Ultimately what occurs in Alice is conscious art married to unconscious science. This is itself predictive, I believe, of what science and art may look like in coming centuries when there will have to be an amalgam, a blend of conscious art and conscious science with unconscious art and unconscious science. Science fiction becoming artistic fact.

What speculations and findings of contemporary science did Carroll anticipate? Certainly we can see chaos theory in the Red Queen's croquet party and the trial of the Knave; relativity theory in Alice's fall down the rabbit hole, in the Mad Hatter's time-warped tea party, and warped space every time Alice eats a piece of mushroom; the quantum effect of observation in the sheep shop; a world ruled by the Red King and the Queen's blind and aimless fury of randomness; as well as wormholes and the rabbit hole to other universes. Numerous other correspondences with modern science are foreshadowed in the *Alice* books, including anti-matter, steriosomers, anti-gravity, black light, non-locality, and faster-than-light, some of which may be found in Carroll's last novel, *Sylvie and Bruno*.[7]

Alice is a pint-sized Queen Victoria whose veneer of manners and propriety hid the Dickensian chaos and jumble of the real Victorian life. Her brain is being stimulated for the dendritic growth required for the 20th century. As these books have served as a major stimulus to ideas, a different way of seeing that grows the brain for present complexity and paradox has emerged. In many ways we can say we live in the world found through the black hole, and you have only to look at the news on television to believe in 'six impossible things before breakfast.'

---

1 Martin Gardner, *The Annotated Alice,* Bramhall House, New York: 1960. We are deeply indebted to Mr. Gardner for his incisive investigation into these correspondences with modern science. Some of which are : pp. 27-28; 99 (relativity theory and the Moebius strip, or projective plane); p. 39 (expanding/shrinking universe); pp. 96-97 (time reversal); pp. 180-184 (asymmetrical objects, anti-gravity, black light, steriosomers, anti-matter), pp. 192-193 (group theory, atomic theory), and numerous other examples to be found throughout Gardner's masterful analysis.

There is a difference between nonsense and absurdity. In nonsense, nothing makes any sense at all. In absurdity you follow the laws of sense until all sense dissipates and you are in a universe of new laws—you undergo a paradigm shift. The writer of the absurd creates an alternative universe. In our case with Carroll, it is one that has bled through with its strange and quirky laws, its loose formulations into our own. We have mathematicized what Carroll has fantasized—but it is a mathematics of Wonderland—which appreciates patterns of possibilities more than objective facts, causal weaves rather than specific causes.

Alice and the Quantum Cat explores these patterns of possibilities by envisioning them for a curious Alice in a modern day 'science-faction-fantasy' experiment, blending conscious art and conscious science with unconscious art and unconscious science.

Quantum physics describes the bizarre and paradoxical reality beyond our senses that is the unseen order of our universe. It is at once the most accurate and complex physical theory ever, predicting the behavior of subatomic particles to twelve decimal places. Yet, perhaps because the wacky world quantum physics describes is counter-intuitive, its secrets remain hidden in a language few but scientists understand.

This book brings together many of the leading interpreters of the New Physics to make tangible the invisible, miraculous mysteries of looking-glass inner world for the non-scientist. Here we enter more imaginary worlds for Alice to inhabit, and perhaps retell some of those other stories that 'lived and died like summer midges, each in its own summer afternoon.'

ACKNOWLEDGMENTS

I am deeply indebted to Fred Alan Wolf and Dianne Collins, who started with me on this odyssey in August 1993, when we spent a week together developing ideas about Quantum Reality for a proposed Public Television series entitled, *Alice in Quantumland*. Failing to receive financial support for the TV project, Fred and Dianne resumed their lives, and as soon as I was financially able, I began commissioning chapters for a book of the same title from Danah Zohar, Nick Herbert, F. David Peat, Amit Goswami, Brian Swimme and Fred Alan Wolf. I am deeply indebted to these fine Minds who have patiently maintained their association with this project over the years.

Science adviser Nick Herbert's brilliant Mind and talented hands are evident throughout this book. Every editor needs an editor, and later the inimitable David Peat stepped in to birth her in English.

I also want to thank the other talented authors and consultants whose astonishing conceptions and insights are contained within: Beverly Rubik, Peter Russell, John P. Briggs, Jean Houston, and my dear, dear friend on the Other Side, Marilyn Ferguson.

And lastly, but perhaps most of all, I would like to thank Lewis Carroll, whose life and magical paradigm-shattering entertainments inspired us, provided the basis for this book, and contributed to its fulfillment in many ways both known and unknown.

W.B.S.

BIOGRAPHIES

WILLIAM BRANDON SHANLEY, CREATOR, EDITOR and CONTRIBU-TOR, is a writer, broadcaster and documentary filmmaker, who lives in New Haven, Connecticut. His programs on American history, espionage, war, conflict resolution, and the great personalities of the 20th century have appeared on PBS, Discovery, A&E, CNN, and been distributed by Turner Home Video and Time Video. William's 1986 documentary feature, *The Made-for-TV Election with Martin Sheen*, demonstrated a 'TV effect' in news coverage in all phases of the Jimmy Carter-Ronald Reagan election of 1980 and heralded the radical rise of show business values in news programming. His on-going passion to understand the reality behind the consensus reality and bring the new worldview to the public, led to the development of the science novel, *Lewis Carroll's Lost Quantum Diaries*, published in Germany and Japan, and this present anthology, *Alice and the Quantum Cat*.

Shanley worked as a writer at CNN following the 9/11 attacks and the early months of the US attack on Afghanistan. These events stimulated his multi-year investigation into hidden networks of power and structures of sin in the public sphere of the United States, *America's Divine Comedy*. The documentary feature reprises Dante Alighieri's 14th-century search for a 'true path' through the Inferno, Purgatory and Paradise. The film will the completed in the autumn of 2011.

The GiveGetShare.net 'gift economy' is being launched in 2011. William developed the free gifts and wishes pool over a number of years with his project partner, Timothy Wilken, MD, who is also a synergy scientist, to supplement the collapsing money system.

In the future, William will focus on a series of films and books to introduce the post-scarcity era he calls 'The Age of Infinity.' This vision was inspired by the author scientists who mentored him and whose creations constitute *Alice and the Quantum Cat*.

NICK HERBERT, PhD, SCIENCE ADVISOR and CONTRIBUTOR, is the author of *Quantum Reality*; *Faster Than Light*; *Elemental Mind*; and a chapbook *Physics on All Fours*. He devised the shortest proof of Bell's Theorem, had a hand in the Quantum No-Cloning Theorem and recently proved that 'A Pair of Quanta Cannot Be Wed.' He has held both industrial and academic posts and for several years ran invitational seminars on Physics Foundations at Esalen Institute. He maintains a blog at http://quantumtantra.blogspot.com and lives in Boulder Creek, CA.

JOHN P. BRIGGS, PhD, is the author of *Trickster Tales*, a collection of stories published by Fine Tooth Press (2005). *Booklist* said of the collection, 'There is a breathless quality to Briggs' fiction that is not so much suspense as a sense of suspension, as though the moorings of reality had been cut.' He is also the author and co-author of nonfiction books on creativity, aesthetics and chaos theory: *Fractals, the Patterns of Chaos; Fire in the Crucible; Seven Life Lessons of Chaos;* and *Turbulent Mirror,* both books with F. David Peat. He is an editor and past senior editor of the national literary journal, *Connecticut Review,* and a CSU Distinguished Professor of Compositional and Relational Aesthetics in the Writing Department at Western Connecticut State University in Danbury.

DIANNE COLLINS, CREATOR and CONTRIBUTOR, is a creative writer and visionary whose passion is showing people new ways of seeing that can enhance the quality of their personal and professional lives. Dianne's study of the evolution of human awareness, her ability to discern what's in people's hearts, and her love of humanity led her to create, along with husband, Alan Collins, the QuantumThink® system of thinking.

For more than 10 years Dianne and Alan have trained executives, managers, consultants and staff in leading companies, as well as entrepreneurs and individuals worldwide, in QuantumThink® principles using a coaching model. A sampling includes Accenture, AT&T, Chase Bank, Dupont, Grainger, IBM, and McKinsey. United States government agencies include the Office of the Assistant Secretary of Defense, the Federal Aviation Administration, the Federal Executive Institute, Office of Personnel Management, the United States Mint, and former Vice-President Al Gore's National Partnership for Reinventing Government.

MARILYN FERGUSON was an author, lecturer, poet and science journalist. Her 1980 best-seller, *The Aquarian Conspiracy*, reported on personal and social transformation and the paradigmatic revolution now underway. The book has appeared in twelve languages and in 1990 was named 'Book of the Decade' by the readers of New Options, a political newsletter.

In 1975, Ferguson turned an interest in human potential into an influential monthly newsletter, *Brain/Mind Bulletin*, which reported on new discoveries in neuroscience and psychology. That work led her to discern that a massive 'cultural realignment' was occurring, a conspiracy in the root sense of disparate forces all breathing together to produce personal and social change.

According to the *Los Angeles Times*, *The Aquarian Conspiracy* was the era's first comprehensive analysis of seemingly unconnected efforts—scientists investigating biofeedback, midwives running alternative birthing centers, politicians encouraging creative government, a Christian evangelist promoting meditation, an astronaut exploring altered states of consciousness—that were 'breathing together' in their break from main-stream Western practices and beliefs in medicine, psychology, spirituality, politics and other fields.

The book's message was optimistic. 'After a dark, violent age, the Pi-scean, we are entering a millennium of love and light—in the words of the popular song "The Age of Aquarius," the time of "the mind's true lib-eration,"' Ferguson wrote. Aquarians, by her definition, were people who sought a revolution in consciousness, to 'leave the prison of our condition-ing, to love, to turn homeward. To conspire with each other and for each other.' As the activities she chronicled moved from the fringe of society to-ward its center, Ferguson was embraced as a beacon. Her book became 'the most commonly accepted statement of Movement ideals and goals,' wrote J. Gordon Melton in the *New Age Encyclopedia*.

Brain/Mind popularized frontier science through special issues and re-ports including Karl Pribram's holographic Model of Reality, Rupert Shel-drake's theory of morphic resonance, and Ilya Prigogine's theory of dissi-pative structures, for which Prigogine won the Nobel Prize for Chemistry in1977. In 1996, Dr. Prigogine's basic reformulation of physics was first published in *Brain/Mind*.

In 1994, Ms. Ferguson was named 'Brain Trainer of the Year' by the American Society of Training and Development, and also received the El-mer & Alyce Green Award for her contribution to scientific understanding, from the International Society for Subtle Energy Medicine. Other awards include the Brandeis University Library Trust Award and an honorary Doc-torate of Letters from John F. Kennedy University.

Ferguson served on the board of directors of the Institute of Noetic Sci-ences, and befriended such diverse figures of influence as inventor and theo-rist Buckminster Fuller, spiritual author Ram Dass, Nobel Prize-winning chemist Ilya Prigogine and billionaire Ted Turner. Ferguson's work also in-fluenced Vice President Al Gore, who participated in her informal network while a senator and later met with her in the White House.

Ferguson died unexpectedly on October 19, 2008.

AMIT GOSWAMI, PhD, is professor emeritus in the physics department of the University of Oregon, Eugene, Oregon where he has served since 1968. He is a pioneer of the new paradigm of science called *science within consciousness*.

Dr. Goswami is the author of the highly successful textbook *Quantum Mechanics* that is used in universities throughout the world. His two vol-ume textbook for nonscientists, *The Physicist's View of Nature* traces the decline and rediscovery of the concept of God within science.

Dr. Goswami has also written many popular books based on his re-search on quantum physics and consciousness. In his seminal book, *The Self-Aware Universe*, he approached the quantum measurement problem by elucidating the famous observer effect while paving the path to a new paradigm of science based on the primacy of consciousness.

Subsequently, in *The Visionary Window*, Goswami demonstrated how science and spirituality could be integrated. In *Physics of the Soul* he developed a theory of survival after death and reincarnation. His book *Quantum Creativity* is a tour de force instruction about how to engage in both outer and inner creativity. *The Quantum Doctor* integrates conventional and alternative medicine.

In his latest book, *God is Not Dead*, Dr. Goswami explores what quantum physics tell us about our origins and how we should live.

In his private life, Dr. Goswami is a practitioner of spirituality and transformation. He calls himself a quantum activist. He appeared in the film *What the Bleep Do We Know?*, *The Dalai Lama Renaissance*, and the award winning documentary *The Quantum Activist*. Dr. Goswami is also a faculty member of Quantum University.

JEAN HOUSTON, PhD, scholar, philosopher and researcher in human capacities, is one of the foremost visionary thinkers and doers of our time, and one of the principal founders of the Human Potential Movement. A powerful and dynamic speaker, she holds conferences and seminars with social leaders, educational institutions and business organizations worldwide.

She is a prolific writer and author of 26 books including *A Passion for the Possible*; *Search for the Beloved*; *Life Force*; *The Possible Human*; *Public Like a Frog*; *A Mythic Life: Learning to Live Our Greater Story*; and *Manual of the Peacemaker*.

In 1965, along with her husband Dr. Robert Masters, Dr. Houston founded The Foundation for Mind Research. She is also the founder and principal teacher since 1982 of the Mystery School, a school of human development, a program of cross-cultural, mythic and spiritual studies, dedicated to teaching history, philosophy, the New Physics, psychology, anthropology, myth and the many dimensions of human potential. She also leads an intensive program in social artistry with leaders coming from all over the world to study with Dr. Houston and her distinguished associates.

As advisor to UNICEF in human and cultural development, she has worked to implement some of their extensive educational and health programs. Since 2003, she has been working with the United Nations Development Program, training leaders in developing countries throughout the world in the new field of social artistry. Dr. Houston has also served for two years in an advisory capacity to President and Mrs. Clinton as well as helping Mrs. Clinton write, *It Takes A Village To Raise A Child*. She has also worked with President and Mrs. Carter and counseled leaders in similar positions in many countries and cultures.

Jean Houston has worked intensively in 40 cultures and 100 countries helping to enhance and deepen their own uniqueness while they become part of the global community. Her ability to inspire and invigorate people enables her to readily convey her vision—the finest possible achievement of the individual potential.

F. DAVID PEAT, PhD, is a physicist with wide interests that include the links between art, science and psychology. In 1996 he moved to the medieval village of Pari in Tuscany where he created the Pari Center for New Learning that hosts courses, conferences and a visitors' program. He has served as a Scientist in Residence at a number of art colleges in England. He is a fellow of the World Academy of Art and Science and a SARChI Distinguished Fellow at the University of South Africa.

Dr. Peat is the author, and co-author, of numerous books including: *Seven Life Lessons of Chaos: Spiritual Wisdom from the Science*; *Infinite Potential: The Life and Times of David Bohm*; *Synchronicity: The Bridge Between Matter and Mind*; *Who's Afraid of Schrodinger's Cat?: All the New Science Ideas You Need to Keep Up With the New Thinking*; *Looking-Glass Universe*; *The Philosopher's Stone: Chaos, Synchronicity and the Hidden Order of the World*; *Superstrings: And the Search for the Theory of Everything*; *Einstein's Moon: Bell's Theorem and the Curious Quest for Quantum Reality*; *Turbulent Mirror: An Illustrated Guide to Chaos Theory and the Science of Wholeness*; *Blackfoot Physics, Science Order and Creativity*, *Seven Live Lessons of Chaos*, *The Blackwinged Night*; *From Certainty to Uncertainty*; *Pathways of Chance*; *Gentle Action*; and *A Flickering Reality: Cinema and the Nature of Reality*. As a broadcaster, Dr. Peat has presented scientific topics on radio and television. For CBC Radio, his credits include a series of twenty one-hour programs, *Physics and Beyond*, as well as the five, one-hour series *Mind and Brain*. He recently featured in the documentary *Turtle Island: Journey with F. David Peat*.

BEVERLY RUBIK, PhD, earned her PhD in biophysics in 1979 at the University of California at Berkeley. She is internationally renowned for her pioneering work in frontier science and medicine. Her main area of focus is research on the subtle energetics of living systems, including the human energy field, energy medicine devices, and whole-person health and healing. She has published over 80 papers and two books. Dr. Rubik presently serves on the editorial boards of *Journal of Alternative & Complementary Medicine*; *Evidence-Based Integrative Medicine*; and *Integrative Medicine Insights*. She has served on the advisory boards of various distinguished organizations, including the Program in Integrative Medicine at University of Arizona under Dr. Andrew Weil.

Dr. Rubik was one of 18 Congressionally-appointed members of the Program Advisory Board to the Office of Alternative Medicine at the US National Institutes of Health (NIH) from 1992-1997, and chaired the NIH panels on bioelectromagnetic medicine and energy healing. This was the precursory organization to National Center for Complementary and Alternative Medicine.

In 1996, Dr. Rubik founded the Institute for Frontier Science (IFS), a nonprofit corporation for research and education. Laurance S. Rockefeller, Sr., helped support the founding of the IFS. In 2002, IFS was awarded an NIH center grant for frontier medicine research on biofield science in consortium with researchers at the University of Arizona. Dr. Rubik was a project director in this consortium and conducted studies on Reiki, a form of Japanese spiritual healing, and on qigong therapy, a healing practice that originated in China. Since then, Dr. Rubik recently completed a project on brainwave measurements and positive affect and is currently conducting research in several areas including distant healing; the subtle properties of drinking water; and a project on sustainable diet and inflammatory markers. She is a core professor in the doctoral programs in Interdisciplinary Studies at Union Institute and University, Cincinnati, OH; and also teaches part-time in Integrative Health at the California Institute of Integral Studies in San Francisco, CA.

Beverly Rubik has won several awards for her research, including the Alyce and Elmer Green award for her pioneering science, awarded to her in 2009 by the International Society for the Study of Subtle Energies and Energy Medicine. She has been interviewed on various television programs, including the most popular morning program in the US, *Good Morning America* (ABC-TV), where she presented her research on the human energy field in December 2000. She serves as a consultant in the health care industry on maverick health and wellness products and as a holistic health practitioner and educator to individual clients.

PETER RUSSELL, MA, DCS, is a world-renowned physicist and philosopher who has published several books on the subject of the deeper, spiritual significance our times. His revolutionary perspective on humanity's place in the universe was explored in his critically-acclaimed book, *The White Hole in Time: Our Future Evolution and the Meaning of Now*. Exploring the patterns behind our long evolutionary journey, as well as the nature of time itself, he shows how our future may culminate in a profound and startlingly positive and creative evolutionary zenith towards which the universe has been building for millions of years. Peter Russell's other books include: *From Science to God: The Meaning of Consciousness and the Meaning of Light*; *The Global Brain*; *The Creative Manager*; *The Brain Book*; *The Upanishads*; and *The TM Technique*. Peter Russell has been hailed as the 'new Buckminster Fuller' and proclaimed a modern visionary by Ted Turner, John Sculley (formerly CEO of Apple Computers), and many others.

BRIAN THOMAS SWIMME, PhD, is a professor of cosmology on the graduate faculty of the California Institute of Integral Studies in San Francisco.

He received his PhD from the department of mathematics at the University of Oregon specializing in gravitational dynamics and singularity theory. Swimme was a faculty member in the Dept. of Mathematics at the University of Puget Sound in Tacoma, Washington from 1978-1981. He was a member of the faculty at the Institute for Culture and Creation Spirituality at Holy Names University in Oakland, California from 1983-1989.

Swimme's primary field of research is the nature of the evolutionary dynamics of the universe. He brings us a meaningful interpretation of the human as an emergent being within the universe and Earth. His central concern is the role of the human within the Earth community. Toward this goal, in 1989, Swimme founded the Center for the Story of the Universe, a production and distribution affiliate of the California Institute of Integral Studies.

His published work includes *The Universe is a Green Dragon*; *The Universe Story* written with Thomas Berry, and *The Hidden Heart of the Cosmos*. Swimme's books have been translated into eight languages. Swimme was featured in the television series *Soul of the Universe* (BBC, 1991) and *The Sacred Balance* produced by David Suzuki (CBC and PBS, 2003). He is the producer of a twelve-part DVD series *Canticle to the Cosmos* which has been distributed worldwide. Other DVD programs featuring Swimme's ideas include *Earth's Imagination* and *The Powers of the Universe*.

FRED ALAN WOLF, PhD, CREATOR and CONTRIBUTOR, aka 'Dr. Quantum,' is a world-renowned physicist, writer, and lecturer who also conducts research on the relationship of quantum physics to consciousness.

In 1963, he earned his Ph.D. in theoretical physics from UCLA and began researching the field of high atmospheric particle behavior following a nuclear explosion. His inquiring mind has delved into the relationship between human consciousness, psychology, physiology, the mystical, and the spiritual. His investigations have taken him from intimate discussions with physicist David Bohm to the magical and mysterious jungles of Peru, from master classes with Nobel Laureate Richard Feynman to the high deserts of Mexico, from a significant meeting with Werner Heisenberg to the hot coals of a firewalk.

Dr. Wolf's work in quantum physics and consciousness is well known through his popular and scientific writing including 13 books, three audio CD courses and numerous film, TV and radio appearances.

He is the author of *Parallel Universes: The Search for Other Worlds*; *The Dreaming Universe*; *The Eagle's Quest: A Physicist Finds the Scientific Truth at the Heart of the Shamanic World*; *The Spiritual Universe*; *Mind into Matter, Matter into Feeling: A New Alchemy of Science and Spirit*; *The Yoga of Time Travel: How the Mind Can Defeat Time*; *Dr. Quantum's Little Book of Big Ideas: Where Science Meets Spirit*; and his latest book, *Time Loops and Space Twists: How God Created the Universe*. He received the National Book Award for *Taking the Quantum Leap: The New Physics for Nonscientists*.

Dr. Wolf has appeared in a number of documentary films including *What the Bleep Do We Know?*; *The Secret*; *The Evidence for Heaven*; *Spirit Space: A Journey Into Your Consciousness*; *Star Trek IV, Special Collector's Edition* discussing 'Time Travel: The art of the possible', and others. He is frequently interviewed on radio and television, and lectured at many prominent organizations and institutions including the Smithsonian Museum in Washington, DC.

Dr. Wolf has also appeared on The Discovery Channel's *The Know Zone*; *Sightings*; *Thinking Allowed*; *The Fabric of Time*; *The Case for Christ's Resurrection*; *Down the Rabbit Hole*; and the PBS series *Closer to Truth*.

DANAH ZOHAR is a management thought leader, physicist, philosopher and author. Her best-selling books include *Spiritual Capital: Wealth We Can Live By*; *SQ: Spiritual Intelligence—The Ultimate Intelligence*, which constitute ground-breaking work on SQ, spiritual intelligence and spiritual capital; *ReWiring the Corporate Brain*; *The Quantum Society*; and *The Quantum Self*, previous work which extends the language and principles of quantum physics into a new understanding of human consciousness, psychology and social organization, particularly the organization of companies. In 1997 she co-authored with her husband Ian Marshall and F. David Peat, *Who's Afraid of Schrödinger's Cat?: All the New Science Ideas You Need to Keep Up with the New Thinking*.

Danah regularly speaks at leadership forums and works with corporate leadership teams worldwide. She has worked with the leadership initiatives of both local and national governments. She established The Oxford Academy of Total Intelligence as an educational centre and consultancy, with a mission to create a sustainable future for society through the development of corporate and organizational leadership, purpose and motivation.

Danah Zohar is currently writing a new book, *If You Know Who You Are: A Journey Into Wholeness and Self-Integration*, which aims to be the culmination of her life's work.

BIBLIOGRAPHY

The following bibliography includes major references for each of the chapters. Biographical information on Charles Lutwidge Dodgson and Lewis Carroll has been culled from numerous biographies, and is used throughout the book. We are particularly indebted to Morton N. Cohen and Martin Gardner, whose masterful works were referenced time and again. Other insights and information may be found in the published works of the authors in their biographies.

W.B.S.
Editor and Creator

Chapter One: Beside the River Bank

Marcin, Raymond B. (2006). *In Search of Schopenhauer's Cat: Arthur Schopenhauer's Quantum-Mystical Theory of Justice*. Washington, DC: Catholic University of America Press.

Chapter Two: Alice Meets Professor Flow

Wolf, Fred Alan Wolf. (1981). *Taking the Quantum Leap*. New York: HarperCollins.

Goswami, Amit. (1995). *The Self-Aware Universe: How Consciousness Creates the Material World*. New York: Tarcher/Putnam.

Zohah, Danah and Marshall, I.N. (1994). *The Quantum Society: Mind, Physics and a New Social Vision*. New York: William Morrow.

Wald, George 'Life and Mind in the Universe' Quantum Biology Symposium, No. 11, *International Journal of Quantum Chemistry*. (1984). Hoboken, NJ: John Wiley & Sons.

Chapter Three: Deep Reality Research

Watts, Alan W. (1962). *The Joyous Cosmology*. New York: Random House.

Herbert, Nick. (1985). *Quantum Reality: Beyond the New Physics*. Garden City, NY: Doubleday Anchor.

Shulgin, Alexander and Ann. (1991). *Pihkal: A Chemical Love Story*. Berkeley, CA: Transform Press.

Chapter Four Oyster Quadrille

Herbert, Nick. (1985). *Quantum Reality: Beyond the New Physics*. Garden City, NY: Doubleday Anchor.

Chapter Five: The Quantum Hussy

Zohar, Danah with Marshall, I. N. (1990). *The Quantum Self: Human Nature and Consciousness Defined by the New Physics*. New York: Quill/William Morrow.

Zohah, Danah and Marshall, I.N. (1994). *The Quantum Society: Mind, Physics and a New Social Vision*. New York: William Morrow.

Chapter Six: One Mind

Goswami, Amit. (1995). *The Self-Aware Universe: How Consciousness Creates the Material World*. New York: Tarcher/Putnam.

Chapter Seven: Alice in Bohm-Land

Bohm, David. (1980). *Wholeness and the Implicate Order*. Boston/London: Routledge & Kegan Paul.

Bohm, David and Peat, F. David. (1987). *Science, Order and Creativity*. New York: Bantam Books.

Lee Nichol. (Ed.). (2002). *The Essential David Bohm*. London: Routledge.

Briggs, John and Peat, F. David. (1984). *Looking Glass Universe: The Emerging Science of Wholeness*. New York: Simon & Schuster.

Chapter Eight: The Center of the Universe

Berry, Thomas and Swimme, Brian. (1994). *The Universe Story: From the Primordial Flaring Forth to the Ecozoic Era: A Celebration of the Unfolding of the Cosmos*. New York: HarperCollins.

Chapter Nine: Chaosland

Briggs, John and Peat, F. David. (1999). *Seven Life Lessons of Chaos: Spiritual Wisdom from the Science of Change*. New York: HarperCollins.

Briggs, John and Peat, F. David. (1990). *Turbulent Mirror: An Illustrated Guide to Chaos Theory and the Science of Wholeness*. New York: Harper and Row.

Chapter Ten: Queen Rosie

Keller, Evelyn Fox. (1986). *Reflections on Gender and Science*. New Haven, CT: Yale University Press

Chapter Eleven: Alice in Ireland

O'Brien, Flann. (1967). *At Swim-two-Birds*. London: Penguin

O'Brien, Flann. (2002). *The Third Policeman*. Champaign, IL/London: Dalkey Archive Press.

Berkeley, George. (2005). *Principles of Human Knowledge*. London: Penguin.

Chapter Twelve: The Looking Glass of Art

Briggs, John and Monaco, Richard. (1990). *Metaphor: The Logic of Poetry.* New York: Pace University Press.

Briggs, John. (1992). *Fractals: The Patterns of Chaos. New York:* Simon and Shuster.

Briggs, John and Peat, F. David. (1990). *Turbulent Mirror: An Illustrated Guide to Chaos Theory and the Science of Wholeness.* New York: Harper and Row.

Chapter Thirteen: Biologyland

Goswami, Amit. (1995) *The Self-Aware Universe: How Consciousness Creates the Material World.* New York: Tarcher/Putnam.

Goswami, Amit. (1997). *Science and Spirituality.* Delhi, India: Munshiram Monoharla.

Chapter Fourteen: Alice in the Biolight Community

Rubik, Beverly. 'Natural Light from Organisms: What, If Anything, Does It Tell Us?' *Noetic Sciences Review.* Summer 1995, Sausalito, CA: Institute of Noetic Sciences, pp. 10-15.

Rubik, Beverly. 'Energy Medicine and the Unifying Concept of Information' *Alternative Therapies in Health and Medicine.* Vol. 1(1), March 1995, pp. 34-39.

Rubik, Beverly. 'Bioelectromagnetics and the Future of Medicine' *Administrative Radiology Journal.* Vol. XVI (8), August 1997, pp. 38-46.

Rubik, Beverly. (1996). *Life at the Edge of Science.* Oakland, CA: Institute for Frontier Science.

Rubik, Beverly. (Ed.). (1992). *The Interrelationship Between Mind and Matter.* Philadelphia, PA: Center for Frontier Sciences at Temple University.

Chapters Fifteen: Underground

Herbert , Nick. (1985). *Quantum Reality: Beyond the New Physics.* Garden City, NY: Doubleday Anchor.

Pagels, Heinz R. (1982). *The Cosmic Code: Quantum Physics as the Language of Nature.* New York: Simon & Schuster.

Deutsch, David. (1997). *The Fabric of Reality: Towards a Theory of Everything.* London: Allen Lane.

Goswami, Amit. (1995) *The Self-Aware Universe: How Consciousness Creates the Material World.* New York: Tarcher/Putnam.

Herbert, Nick. (1994, reprint edn.). *Elemental Mind: Human Consciousness and the New Physics.* New York: Plume.

Radin, Dean. (1997). *The Conscious Universe.* New York: HarperCollins.

Chapter Sixteen: A Photon's Eye View
Various. See Peter Russell's biography.

Chapter Seventeen Caught in the Web of Time
Wolf, Fred Alan. (1996). *The Spiritual Universe: How Quantum Physics Proves the Existence of the Soul.* New York: Simon & Schuster.

Chapter Eighteen: Universal Attraction
Swimme, Brian. (1984). *The Universe is a Green Dragon: A Cosmic Creation Story.* Rochester, VT: Bear & Co.

Chapter Nineteen The Symphony of Creation
The Diary of Virginia Woolf. Vols. 1-5. (1979-85). New York: Mariner Books.
The Letters of Virginia Woolf. Vols. 1-6. (1975-80). London: The Hogarth Press.

Chapter Twenty. Alice Rides the Photon
Young, Arthur M. (1976) *The Reflexive Universe: Evolution of Consciousness.* Cambria, CA: Anodos Foundation.

SCIENTIFIC GLOSSARY

Nick Herbert

QUANTUM THEORY—Our modern theory of light and Matter: both described the same way—as possibility waves when not looked at, as actual particles when observed. Put in mathematical form by Werner Heisenberg, Erwin Schrödinger and Paul Dirac in the late 1920s.

QUANTUM—Latin for 'how much.' Refers to the particle nature of Matter/ light. To 'quantize' something is to divide it into discrete bits.

PLANCK'S CONSTANT—a fundamental constant of Nature that sets the scale for Quantum activity. Planck's constant of action is a measure of 'the least thing that can happen,' the lowest denomination or basic coin of the Quantum Realm. Princeton's physics describes everyday phenomena so well because Planck's constant is exceedingly small. Introduced into physics by John Planck in 1900.

ON—Greek for 'entity.' Suffix used by physicists to designate the particle or quantized aspect of phenomena. Examples: particles of light are 'photons'; 'phonons' are quantized sound waves; protons, neutrons, muons, pions, leptons—the list is endless. An electron is a Quantum of electricity.

WAVE—a form of energy that is spread out over space, can freely interpenetrate other waves, characterized by its frequency, wavelength and velocity.

PARTICLE—a form of energy concentrated in space, impenetrable by other particles, characterized by its mass, spin and momentum.

UNCERTAINTY PRINCIPLE—a characteristic feature of Quantum theory is that all Quantum attributes occur in complementary pairs. If one attribute of the pair is precisely observed, the partner attribute becomes imprecise. Examples of paired Quantum attributes include momentum/position, energy/time and Xspin/Yspin.

WAVE FUNCTION—In Quantum theory, every physical system is 'represented by' a mathematical expression called 'its wave function' from which the results of all conceivable measurements on that system can be calculated. One of the landmark achievements of the twentieth century was the calculation of the wave function of the hydrogen atom which allows the wavelength of every spectral line produced by glowing hydrogen gas to be precisely predicted. Fred Alan Wolf calls the wave function 'the QWIFF!'

WAVE FUNCTION COLLAPSE–An isolated Quantum system is completely described by a spread-out vibrating entity called 'its wave function.' However whenever we observe a system it looks like a compact particle (called a 'Quantum'). The (presumed and utterly mysterious) transition between wave description and particle actuality has been dubbed 'the collapse of the wave function.' Fred Wolf calls this collapse 'popping the QWIFF!'

EIGENFUNCTION–For a particular Quantum system, almost every kind of observation you can make will give an unpredictable result. However every isolated system possesses a few special observations such that if those observations are made the results are completely predictable. The wave function that describes the system is called an EIGENFUNCTION of those (special) observables and the (completely predictable) results are called EIGENVALUES.

SCHRÖDINGER'S EQUATION–One way of calculating the wave function devised by Erwin Schrödinger in 1926. A year earlier Werner Heisenberg had derived Matrix Mechanics and in the 1950s Richard Feynman invented a third way of doing Quantum theory called SUM-OVER-HISTORIES symbolized by his famous FEYNMAN DIAGRAMS.

DIRAC EQUATION–a version of the Schrödinger Equation that incorporates Einstein's Theory of Relativity. One of Dirac's first successful predictions was the existence of antiparticles—particles whose intrinsic properties—charge and spin, for instance—are exact opposites of the properties of ordinary particles. The first antiparticle to be discovered was the positron, the antiparticle of the electron. When a particle meets its anti-particle they annihilate one another producing (usually) a pair of high-energy photons.

SCHRÖDINGER'S CAT–Refers to a famous thought experiment invented by Erwin Schrödinger in 1935 to illustrate the absurdity of Quantum theory's insistence that the world is a wave until observed. Using standard Quantum theoretical ideas Schrödinger showed how one could construct an experiment in which an unobserved cat becomes both alive and dead at the same time. More than half a century later the status of Schrödinger's famous cat is still hotly debated in physics circles.

HILBERT SPACE–An extremely abstract way of describing the Quantum world due to John von Neumann and named after his teacher, mathematician David Hilbert. Hilbert space, roughly speaking, has as many dimensions as there are possible experimental results. The Hilbert space of the simplest spinless point particle has an infinite number of dimensions, each dimension corresponding to one of the possible positions in space where the particle might be observed to be located. A physical system is represented by a direction in Hilbert space called the STATE VECTOR. The components of the state vector along the position dimensions of Hilbert space yield the WAVE FUNCTION.

RELATIVITY THEORY–Our modern theory of space and time devised by Albert Einstein in 1905. Shows that observers moving at different velocities will obtain different results for length and timing measurements on the same phenomenon. Thus space and time are 'relative' for observers in different states of motion. Einstein also showed how to calculate certain Absolutes—the same for all observers—which include the speed of light and the 'invariant space-time interval.' Any physical theory that is not constructed out of Einsteinian Absolute quantities cannot be a correct theory of Nature because it will produce different laws for different observers.

SPEED OF LIGHT–an Einsteinian Absolute, the same measured numerical value for everyone no Matter how they are moving. Einstein's relativity theory shows that no signals can be exchanged nor Matter moved faster than this velocity because such behavior would amount to information or things moving backwards in time in some reference frames.

SPACE-TIME–An Einsteinian Absolute is a kind of mathematical object in a four-dimensional Realm—space-time—consisting of three space dimensions and one time dimension. The entire history of any object is pictured in this 4-D space-time as a curved ribbon called a WORLDLINE.

GENERAL RELATIVITY–Einstein's theory of gravity expressed as a curvature in space-time. (Time is curved too!). General relativity in a nutshell: Mass tells space-time how to curve; space-time tells mass how to move.

BLACK HOLE–a region of space-time so bent by gravity that nothing can escape its pull—not even light. Black holes are almost certainly responsible for certain high-energy astronomical objects and may exist at the center of many galaxies, including our own.

WORMHOLE–a conjectured tunnel made of twisted space-time that directly links one historical situation with another. An alleged shortcut in space-time allowing effectively faster-than-light travel.

BELL'S THEOREM–A mathematical proof by Irish physicist John Stewart Bell that two Quantum particles that have once interacted but are now separate still possess a lingering voodoo-like connection such that action on one particle affects the other in a manner unmediated, unmitigated and immediate. This instant connection between Quantum particles, although certainly faster-than-light, does not violate Einstein's famous speed limit because Quantum randomness scrambles any attempt to exploit the Bell connection for superluminal signaling. Nature uses the superfast Bell link to accomplish Her everyday miracles but humans are barred from this underworld channel by Her intrinsically unbreakable encryption.

QUANTUM ENTANGLEMENT–When two Quantum systems interact, their Quantum representation, except in very unusual cases, describes them as a single inseparable whole, their attributes intermingled in such a manner that what is done to one system instantly alters the attributes of the other. Bell's Theorem proves that this intimate entanglement present in the system's mathematical description is also present in Reality Itself.

CLASSICAL PHYSICS–an immensely successful theory of the motion of ordinary objects invented by Isaac Newton. Newton's world is made of particles whose motion is determined by force fields (such as gravity, electricity and magnetism). Space and time are absolute—the same for all observers. The Newtonian world is deterministic—completely predictable from initial conditions and has been compared to a giant clockwork. Newton's physics is still a very good approximation for large objects moving slowly compared to light speed but Newton's worldview has been superseded by relativity and Quantum theory.

LOCALITY–The notion that a change here can affect a body there only if 'something' passes from here to there. Classical physics is a local theory—its influences are mediated by fields (electric, magnetic or gravity, for instance). If I change a body here, all the fields connected with that body change too and these changes are transmitted to other bodies by a wave that travels through the field at a speed no greater than that of light.

NON-LOCALITY–In Quantum theory changes here are transmitted there also via the mediation of fields (hence LOCAL) but QUANTUM ENTANGLEMENT (which Schrödinger called THE distinctive feature of Quantum theory) allows influences to jump instantly from one place to another (even backwards in time!) without passing through the intervening space. Thus Quantum theory is NON-LOCAL and so (John Bell proved this) is Quantum Reality.

QUANTUM VACUUM–One of the most curious features of Quantum theory is that a region of space empty of all particles and fields still possesses an immense quantity of unobservable energy. The Quantum vacuum is not empty but full of 'virtual particles' of all types and energies that exert subtle changes on Real atoms and molecules located in so-called 'empty space'.

GRAND UNIFIED THEORY (GUT)–The attempt to unite all of Nature's forces into one unified description. James Clark Maxwell unified electric and magnetic forces in the 19th century into one electromagnetic (EM) field. Recently Weinberg, Glashow & Salam united EM with the 'weak force' and progress is being made to bring the 'strong force' and gravity under one umbrella.

MODERN PHYSICS, also NEW PHYSICS–relativity theory and Quantum theory.

COHERENT BIOPHOTONS–a conjecture partly supported by experiment that living cells are linked by a communication system mediated by weak in-phase light.

IMAGINAL WORLD–the Realm from which our mental concepts arise. Many thinkers, from Plato to C.G. Jung believed that this Realm is as Real if not more Real than the world of direct experience.

PERENNIAL PHILOSOPHY–The core teaching of mystics everywhere: that a timeless Divine Reality permeates the universe and that our own inner lives can be viewed as abbreviated experiences of this Divine Ground of Being.

CHAOS THEORY–A branch of Newtonian mechanics that deals with systems whose motion is deterministic but unpredictable due to the impossibility of exactly specifying initial conditions or accounting precisely for external influences. Most natural systems are chaotic—theoretically predictable but random in practice.

STRANGE ATTRACTOR–Although it is impossible to exactly predict the motion of a chaotic system, one can often classify different types of chaos (similar to classifying types of cloud formations). 'Strange attractor' is the class which contains the most complex types of chaotic motion.

TENTATIVE MODELS OF THE SUBATOMIC WORLD

Nick Herbert

1. The Duplex World of Werner Heisenberg. Heisenberg, in 1926 discovered the first mathematical theory that accurately predicted the Quantum facts. Heisenberg realized that the math worked, but the language was a serious problem. He said atoms cannot be described in ordinary language as things.

Instead, Quantum theory represents the unobserved world as possibility waves, meaning the world might really exist, when not looked at, as mere waves of possibility. 'Atoms and elementary particles are not as real [as the phenomenal world]; they form a world of potentialities and possibilities rather than one of things and facts... The probability wave...means a tendency for something. It's a quantitative version of the old concept of potentia in Aristotle's philosophy. It introduces something standing in the middle between an idea of an event and an actual event, a strange kind of physical Reality just in the middle between possibility and Reality.'

2. There is No Deep, Hidden Reality. Danish physicist Niels Bohr's Quantum Reality #1 argues that there is no deep, hidden Reality—only phenomena are 'Real.'

Atoms themselves are not so Real. We know them only indirectly from the results of measurements. Their mode of existence is of a type that can never be grasped by human beings who are constrained to live and think exclusively in the world of phenomena. Both Bohr and Heisenberg stressed the difficulty of describing Quantum Reality in ordinary language. Bohr believed that it was futile to try while Heisenberg tried to conceptualize this invisible Reality as 'possibilities'.

3. Reality is Created by Observation. Quantum Reality #3 argues that 'no phenomenon is Real until it is an observed phenomenon,' echoing Bishop Berkeley's slogan 'Esse Est Percepi' (To be is to be perceived.). In Berkeley's view, God's constant attention keeps the world in existence whenever mortal eyes close. In the Copenhagen view when no one looks the world simply does not exist in a form that humans can conceptualize.

Quantum Realities 2 & 3 make up what is known as the Copenhagen Interpretation.

4. Reality is an undivided wholeness, contrary to the Newtonian view that the world consists of a collection of isolated particles interacting by 'local force fields' such as gravity, electrical and magnetic fields. Erwin Schrödinger's mathematical Quantum possibility wave creates (in theory) immediate, non-local interactions, and John Bell proved that once any two atoms have interacted they remain connected (in Reality) no matter how far apart they may be. David Bohm's 'undivided wholeness' puts Bell's discovery into an explicit mathematical picture.

5. Quantum Logic. Could it be that the Quantum dilemma might be solved by making a radical change in our very laws of thought? Perhaps the world Really consists of atoms whose positions are always definite (hence no 'measurement problem'), but we can only properly talk about these atomic positions using non-Boolean logic, requiring new grammatical rules for combining the words 'and,' 'or' and 'not'. In a world consisting of cows and horses that can be either black and white, you might let all NOT HORSES into a stall and then release all these animals that are NOT BLACK into a second stall. If the sorting of animals obeyed a quantum logic analogous to the polarization sorting of photons you would find that half of the animals in the second stall were white horses.

6. Hugh Everett's Many Worlds Interpretation. This says the universe is made up of strands of sub universes, with each strand a different possible history of what we would call the whole universe. This Quantum formulation reduces the role of the 'observer' and immensely enlarges our view of what the word universe might mean: that the unobserved atom's Quantum possible positions are in fact actualities, not possibilities. The atom is actually in many places at the same time, but each of these atomic positions is located in a different universe.

7. Consciousness Creates Reality. John von Neumann believed that the theory of the Quantum unequivocally requires that it be described in possibility waves—not as a collection of actual objects possessing at all times a definite value for each of their physical attributes. A totally Quantum world is a world of pure possibility in infinitely dimensioned Hilbert Space, where nothing ever actually happens, but hesitates forever on the brink of actuality. In contrast to the old-fashioned, definite 'yes and no' world of Newtonian physics, the Quantum world resembles a fairy-tale land built solely of ambiguous 'maybes.' The world remains in a state of pure possibility except when some conscious Mind decides to promote a portion of the world from its usual state of indefiniteness into a condition of actual existence.

8. NeoRealism. The approach of Louis de Broglie and David Bohm. Ordinary particles behave in a classical, Newtonian sense, but are moved around by entities called 'pilot waves'—each pilot wave attached to its own particle. This theory solves the measurement problem by imagining that atoms and other Quantum entities always have definite positions whether observed or not. Each pilot wave guides its particle by supplying force to it like gravity as well as information like the position of every other particle, then guides its particle on its journey through space via a personal radar wave. These pilot waves are invisible and therefore unobservable. Because they are unseeable, pilot waves are allowed to travel faster than light—as any deep Reality must according to Bell's theorem—without violating Einstein's prohibition against overt superluminal velocities.

QUANTUM REALITY'S TWO UNANSWERED QUESTIONS:

Quantum theory describes the world as waves of possibility when you don't look and as actual particles when you do look. Two very basic and still unanswered questions are:

1. What is really going on in the world before an observation?

Called the 'interpretation question' because it essentially asks what interpretation shall we give (in Reality) to the (theoretical) possibility waves that physicists use to represent the unobserved world?

2. What is a Quantum observation?

Called the 'measurement problem' because it asks which encounters in the world are merely interactions (preserving the possibility structure of Reality) and which encounters are 'measurements,' turning possibility waves into actual particles?

Pari Publishing is an independent publishing company, based in a medieval Italian village. Our books appeal to a broad readership and focus on innovative ideas and approaches from new and established authors who are experts in their fields. We publish books in the areas of science, society, psychology, and the arts.

Our books are available at all good bookstores or online at
www.paripublishing.com

If you would like to add your name to our email list to receive information about our forthcoming titles and our online newsletter please contact us at
newsletter@paripublishing.com

Visit us at **www.paripublishing.com**

Pari Publishing Sas
Via Tozzi, 7
58045 Pari (GR)
Italy

Email: info@paripublishing.com